XIII International Scientific Conference "Tribology for Mechanical Engineering" (TriboMash 2020)

Held online due to COVID-19

IOP Conference Series: Materials Science and Engineering
Volume 996

Moscow, Russia
14 - 16 October 2020

ISBN: 978-1-7138-3303-1
ISSN: 1757-8981

Printed from e-media with permission by:

Curran Associates, Inc.
57 Morehouse Lane
Red Hook, NY 12571

CURRAN ASSOCIATES INC. proceedings.com

Some format issues inherent in the e-media version may also appear in this print version.

Institute of Physics
Dirac House, Temple Back
Bristol BS1 6BE UK

Phone: 44 1 17 929 7481
Fax: 44 1 17 920 0979

techtracking@iop.org

Additional copies of this publication are available from:

Curran Associates, Inc.
57 Morehouse Lane
Red Hook, NY 12571 USA
Phone: 845-758-0400
Fax: 845-758-2633
Email: curran@proceedings.com
Web: www.proceedings.com

TABLE OF CONTENTS

IOP Conf. Series: Materials Science and Engineering **996** (2020) 011001 doi:10.1088/1757-899X/996/1/011001

XIII International Scientific Conference "Tribology for Mechanical Engineering" (TriboMash 2020), 14-16 October 2020, Russia, Moscow

Blagonravov Mechanical Engineering Research Institute of the Russian Academy of Sciences holds a scientific conference «Tribology for Mechanical Engineering" every two years since 1982. XIII international conference «Tribology for Mechanical Engineering" was held on October 14-16, 2020 in Moscow, Russia. Purposes of the conference include discussion of current theoretical and applied problems of friction, wear and lubricant, generalization of the latest results of advanced calculation and experimental researches in relation to work of various branches of mechanical engineering (transport, machine-tool construction, space, oil and gas, etc.), the analysis of researches in the nanotribology, development of new highly effective methods of a tribomonitoring and creation new highly effective antifrictional, wear-resistant materials and lubricants.
In 2020, the conference was held virtually due to restrictions related to COVID19 using the ZOOM platform. The organizers were located at the Blagonravov Mechanical Engineering Research Institute of the Russian Academy of Sciences, Moscow, Russia.
The meetings were held in the format of a speech with an oral presentation. For each participant was allotted 15 minutes including questions. Plenary sessions were held on the first day of the conference.
93 scientists from various cities of Russia and Belarus participated in the conference in 2020. The organizing committee selected 26 articles of the participants for publication in the collection of works.
The conference passed without major problems. Organizing committee connected many participants virtually. The exhibition part of the conference was impossible due to the remote presentation. This negatively affects the development of the experimental part of the tribological research.

List of Conference Committee

Ganiev R. F., Conference Chairman, academician of RAS, Scientific Director of IMASH RAN, Mechanical Engineering Research Institute of the Russian Academy of Sciences (IMASH RAN), Moscow, Russia

Goryacheva I.G., Chairman of Program Committee, academician of RAS, Chairman Of the interdepartmental scientific Council on tribology, Moscow, Russia

Glazunov V. A., Chairman of Organizing Committee, Ph.D., Director of IMASH RAN, Mechanical Engineering Research Institute of the Russian Academy of Sciences (IMASH RAN), Moscow, Russia

Albagachiev A.Yu., Deputy Chairman of Organizing Committee, Mechanical Engineering Research Institute of the Russian Academy of Sciences (IMASH RAN), Moscow, Russia

Buyanovsky I. A., Deputy Chairman of Organizing Committee, Mechanical Engineering Research Institute of the Russian Academy of Sciences (IMASH RAN), Moscow, Russia

Prozhega M.V., Scientific Secretary of the Conference, Mechanical Engineering Research Institute of the Russian Academy of Sciences (IMASH RAN), Moscow, Russia

Samusenko V.D., Scientific Secretary of the Conference, Mechanical Engineering Research Institute of the Russian Academy of Sciences (IMASH RAN), Moscow, Russia

Khrushchev M.M., Mechanical Engineering Research Institute of the Russian Academy of Sciences (IMASH RAN), Moscow, Russia

Yudkin V. F., Mechanical Engineering Research Institute of the Russian Academy of Sciences (IMASH RAN), Moscow, Russia

Gadolina I.V., Mechanical Engineering Research Institute of the Russian Academy of Sciences (IMASH RAN), Moscow, Russia

IOP Conf. Series: Materials Science and Engineering **996** (2020) 011001 doi:10.1088/1757-899X/996/1/011001

Hasyanova D.U., Mechanical Engineering Research Institute of the Russian Academy of Sciences (IMASH RAN), Moscow, Russia

Bolotov A.N., Tver State Technical University, Tver, Russia

Bronovets M.A., Ishlinsky Institute for Problems in Mechanics RAS, Moscow, Russia

Zakharov S.M., Railway research institute of JSC Russian Railways, Moscow, Russia

Kirichek A.V., Bryansk State Technical University, Bryansk, Russia

Kolesnikov V.I., academician of RAS, Rostov-on-Don, Russia

Krasnov A.P., A.N.Nesmeyanov Institute of Organoelement Compounds of Russian Academy of Sciences

(INEOS RAS) Moscow, Russia

Levchenko V.A., Lomonosov Moscow State University, Moscow, Russia

Lyubinin I.A., LLC "RN-Lubricants", Moscow, Russia

Myshkin N.T., V.A. Belyi Metal-Polymer Research Institute of National Academy of Sciences of Belarus. (MPRI NASB), Gomel, Belarus

Parenago O.P., A.V.Topchiev Institute of Petrochemical Synthesis, RAS (TIPS RAS), Moscow, Russia

Popov V.L., Berlin Technical University, Berlin, Germany, Berlin, Germany

Savin L.A., Oryol State University named after I.S. Turgenev, Oryol, Russia

Tsukanov I.Yu., Ishlinsky Institute for Problems in Mechanics RAS, Moscow, Russia

IOP Conf. Series: Materials Science and Engineering **996** (2020) 011002 doi:10.1088/1757-899X/996/1/011002

Peer review declaration

All papers published in this volume of *IOP Conference Journal: Materials Science and Engineering (MSE)* have been peer reviewed through processes administered by the proceedings Editors. Reviews were conducted by expert referees to the professional and scientific standards expected of a proceedings journal published by IOP Publishing.

- **Type of peer review: Open**

For Peer review of submissions was used open system. Based on the results of the speeches, the editors of the Conference selected the most interesting scientific reports and proposed to publish the article in the Proceedings.

- **Conference submission management system:**

The organising committee's email address was used to submit applications for the conference.

- **Number of submissions received:**

The organising committee received 93 submissions.

- **Number of submissions sent for review:**

76 submissions were sent for review.

- **Number of submissions accepted:**

26 submissions were accepted by Editors.

- **Acceptance Rate (Number of Submissions Accepted / Number of Submissions Received X 100):**

Accepted Rate is 30.

- **Average number of reviews per paper:**

One review for paper was prepared.

- **Total number of reviewers involved:**

24 reviewers worked with applications.

- **Any additional info on review process:**

- **Contact person for queries:**
Ph.D. Maksim Prozhega
Bardina, 4, 119334, Moscow, Russia
prmaksim@gmail.com

Laser frictional area treatment of zirconium crystals

V V Alisin[1], M A Borik[2], E E Lomonova[2] and N Yu Tabachkova[2,3]

[1] Mechanical Engineering Research Institute of The Russian Academy of Sciences, Maly Kharitonevsky Lane, 4, Moscow, Russia
[2] Prokhorov General Physics Institute RAS, Vavilov str, 38, Moscow, Russia
[3] National University of Science and Technology "MISIS", Moscow, Russia

vva-imash@yandex.ru

Abstract. The issues of laser treatment of nanostructured partially stabilized crystals of zirconium dioxide are studied in the article. The phenomena of formation of the topography of the surface of crystals after treatment by focused radiation of a repetitively pulsed laser at a wavelength of 1.06 μm are investigated. Structural changes in the surface layer have been studied by X-ray phase analysis. The formation of a golden film on the surface of the treated crystal and ceramic sample of the same chemical composition and the effect of reducing the coefficient of friction have been established. The tribological properties of PSZ crystals are investigated during friction without lubrication in the sliding mode.

1. Introduction

Heating a surface by exposure to laser radiation is a widespread technology used to improve the tribological properties of a surface by hardening or changing the structure of the surface layer of the material being treated. In addition to thermal effects in the interaction of a laser beam with the surface of a solid, others also take place and find application. Various types of continuous and repetitively pulsed lasers of various wavelengths are used for surface modification. Repetitively pulsed lasers are used [1] for discrete hardening of the friction surface. Laser remelting of surfaces of various materials is widely used [2, 3, 4] to change the structure in order to increase the wear resistance of friction surfaces, for example, to obtain [5] self-lubricating boride layers, for heat treatment and laser building-up of wear-resistant coatings [6, 7], to improve tribological properties [8] under conditions of friction without lubrication at high temperatures.

Alloying with boron and solid lubricant CaF_2 was used by laser infusion of the coating to create the effect of self-lubrication [9]. Laser alloying with 10% manganese was carried out to increase the wear resistance and hardness of the surface [10]. The mechanism of material removal and the formation of the surface relief under the action of a laser is described in [11, 12]. As applied to tribological problems, the use of ceramics and crystals based on zirconium dioxide is promising because these materials have a higher fracture toughness and a tendency to adhesion in comparison with most of the known ceramics and crystals. The highest mechanical characteristics are possessed by nanostructured partially stabilized crystals of zirconium dioxide (PSZ crystals) [13]. Furnace heat treatment of crystals of partially stabilized zirconium (PSZ crystals) was studied in [14]. The interaction of laser radiation with the surface of PSZ crystals is an insufficiently explored sphere.

Objectives and aims of the research. Examine the impact of high-power laser radiation on the surface layers of a promising refractory, chemically inert, high-strength, crack-resistant, nanocrystalline crystal of partially stabilized zirconium dioxide in order to increase its tribotechnical properties.

2. Materials and research methods

The object of the study were crystal samples ZrO_2 + 3 mol% Y_2O_3 and Y_2O_3 + 3.5 mol% Y_2O_3 obtained by directional controlled solidification from the melt [13]. In tribological tests, a disc made of У10А steel (chemical composition in % C 0.96 -1.034; Si 0.17 – 0.33; Mn 0.17 – 0.28; Ni 0.2; S 0.018; P 0.025; Cr 0.2; Cu 0.2; Fe 0.07), hardness HRC 49-52 was used as a counter - body.

Laser treatment in a repetitively pulsed mode was carried out on a Kvant-12 setup with focused laser radiation at a wavelength of 1.06 μm. The focusing lens has a focal length of 100 mm and provides a change in the diameter of the light spot in the range of 0.5 - 1 mm; an elliptical spot is formed with the help of attached cylindrical lenses. The experiments were carried out with a light spot 1 mm in diameter, the duration of the radiation pulses was 4 ms, and the repetition rate of the radiation pulses from the setup was 1 Hz. The end surfaces of rod samples of PSZ crystals with a square cross section of 5x5 mm, formed in packages of 5 pieces, were treated with a laser beam by applying parallel tracks to the samples installed in a line.

The phase composition of the materials under study was determined by the method of X-ray diffractometry on a Bruker D8 setup with symmetric recording Θ - 2Θ with a scanning point detector, in the geometry of a parallel beam. Samples for phase analysis were cut from crystals perpendicular to the <100> direction. Since the intensity of reflections from single crystals is much higher than from polycrystals, on the (100) plane, in the presence of several phases in the crystal, on one cut, one can observe simultaneously reflections from different phases, which were separated at large angles 2Θ~130°. The phase fraction was determined from the intensity of the diffraction maxima, which was normalized to the integral reflection coefficient for different phases.

The study of the microstructure of the surface of the samples after laser treatment was carried out on a JEOL 5910 LV scanning electron microscopy unit and an INCA-300 OXFORD X-ray spectrum analyzer with a digital camera.

A quantitative study of the crystallographic orientation, phase composition, and texture of the resulting films after laser treatment at the microlevel was carried out using an electron microscope by electron backscatter diffraction (EBSD). The information obtained from the surface of the object was used; the penetration depth was ~ tens of nanometers.

Tribological tests were carried out on a UMT-1 friction machine on model samples according to the pin-on-disk scheme [15].

3. Results and its discussion

The studied crystals contained two tetragonal phases, transformable (t) and nontransformable (t'), differing in the degree of tetragonality ($c \cdot (2a)^{-0.5}$) 1.014 and 1.005, respectively. The quantitative ratio of phases t/t' varied slightly depending on the concentration of yttrium oxide from 85 (5) / 15 (5) to 80 (5) / 20 (5).

In the experiments, the radiation energy increased stepwise from 1 to 5 J until the formation of a golden film (coating) on the surface of the PSZ crystal, the nature of which is not clear at this stage of work. The film first appeared in the center of the light spot and, with increasing power, completely covered the spot; at the same time, a crater was formed in the center of the spot due to the evaporation of material from the treated surface. To form the next deposit track, the sample was displaced perpendicular to the velocity vector by a distance equal to the diameter of the light spot. In experiments, a stable effect of the formation of a golden film is noted. The surface morphology of PSZ crystals subjected to laser modification is shown in figure 1.

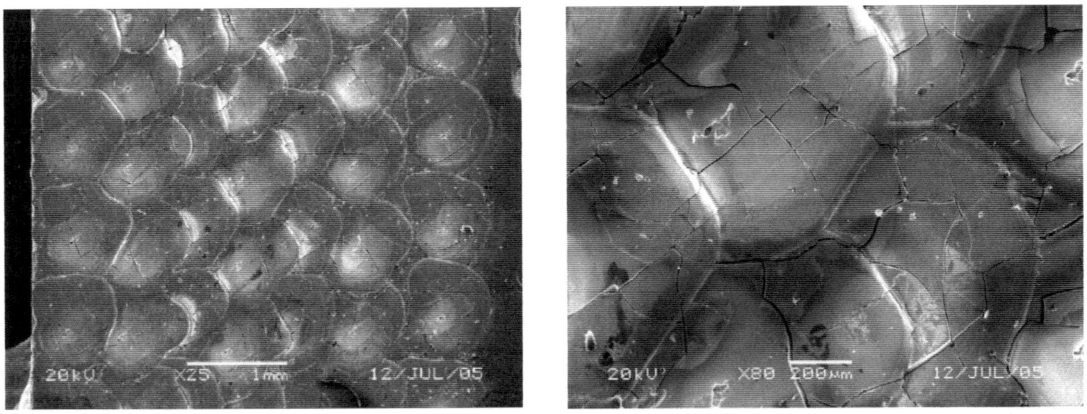

Figure 1. The surface morphology of PSZ crystals of composition ZrO_2 –3 mol% Y_2O_3 exposed to laser radiation, obtained with an electron microscope (x 25).

It is seen that as a result of laser treatment, the crystal surface melted. In general, the modified surface is a fairly regular array of buildup, separated by parapets (bridges). A greater increase in the microscope showed that a film is formed on the surface, some areas of which contain cracks and buildup. On the surface of the PSZ crystal, holes are visible, formed due to the evaporation of the material under the influence of a laser beam. The surface is covered with a network of micro-cracks due to residual tensile stresses following structural changes. The probable mechanism of crack formation is the difference in the coefficients of linear thermal expansion of the PSZ crystal in the base and in the surface layer. Figure 2 shows cracks covering almost the entire treated surface and where the surface composition is controlled.

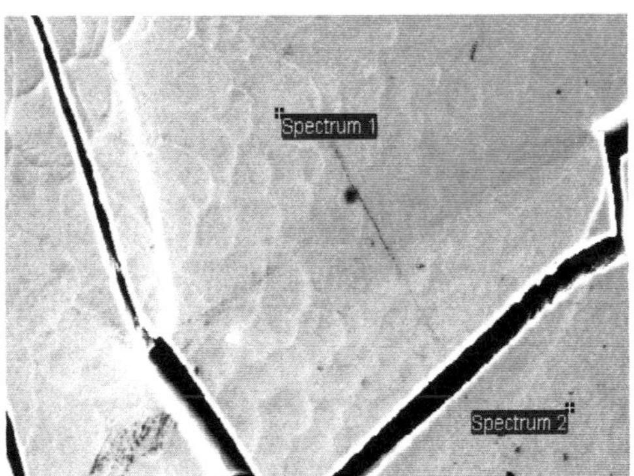

Figure 2. Surface morphology of PSZ crystals of composition ZrO_2 –3 mol% Y_2O_3 exposed to laser radiation. (Enlargement x1000)

Using Energy-dispersive X-ray spectroscopy analysis (EDXR) methods, the chemical composition of the coating formed on samples of PSZ crystals of the composition ZrO_2 –3 mol% Y_2O_3 subjected to laser surface modification was determined. Analysis of the chemical composition of the surface layer of the sample modified by laser treatment (figure 3) showed that the surface is represented by zirconium oxide. It should be noted that there is a deviation from stoichiometry for oxygen and a significant variation of this nonstoichiometry at different dots: from 29.12 to 48.16 for zirconium and from 51.84 to 70.44 for oxygen (data are given in atomic percentages).

The study of the obtained film was carried out after laser treatment of the sample surface by electron-backscattered diffraction (EBSD). The figure shows an image of a section of the sample surface obtained with an electron microscope (figure 3 (a)), and the phase map corresponding to this area (figure 3 (b)).

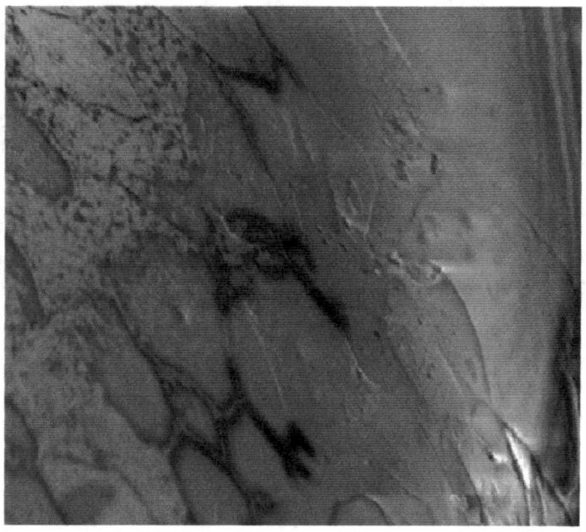

(a)

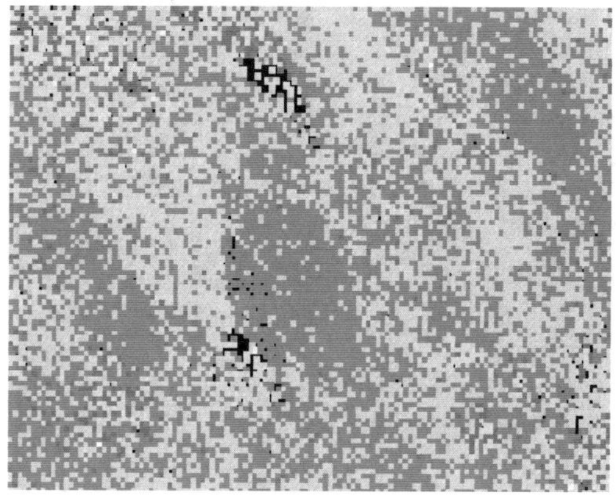

(b)

Figure 3. Image of the sample surface (a), and the phase map (b).

The results obtained indicate that, when exposed to laser radiation, melting occurs on the surface, which leads to a significant change in the phase composition. The phase composition of the film formed after laser treatment consists of tetragonal (pink color), cubic (green), and rhombohedral (blue) phases (figure 3 (b)). A small amount of a monoclinic phase is present. The formation of the phase of metallic zirconium is practically absent; its presence can be regarded as a measurement error. Attention is drawn to the formation of a significant fraction of the orthorhombic phase.

A measure of the antifriction properties is the coefficient of friction, which was determined on model samples according to the pin-on-disk scheme. Samples (pins) are made from the same PSZ crystal. The results of comparative tests of PSZ crystals in the initial state and after laser treatment with pulses with an energy of 3 J and polycrystalline diamond at various pressures at the contact are given in table 1.

Table 1. Coefficients of friction of surfaces during friction of PSZ crystals on steel.

Pressure P, MPa	2.5	5	10	20	30
PSZ crystal without coating	0.45	0.37	0.31	0.28	0.28
PSZ crystal with coating	0.30	0.29	0.27	0.25	0.26
Diamond	-	-	0.26	0.30	0.28

It has been experimentally established that after laser treatment of the surface of PSZ crystals, there is a noticeable (up to 1.5 times) decrease in the region of low pressures of the friction coefficient in the presence of a coating on the friction surface of PSZ crystals.

The results of tribotechnical tests of PSZ crystals after laser treatment are given in tables 2 and 3. It has been established that during friction without lubrication on У10A steel in the contact pressure range of 2.5 - 40 MPa, the wear rate of PSZ crystals after laser treatment varies from $2.42 \cdot 10^{-8}$ to $2.81 \cdot 10^{-7}$. The coefficient of friction gradually decreases as the contact pressure increases from 0.32 to 0.19.

IOP Conf. Series: Materials Science and Engineering **996** (2020) 012001 doi:10.1088/1757-899X/996/1/012001

Table 2. Results of testing samples of ZrO_2 + 3.5 mol.% Y_2O_3 crystals after laser treatment.

p, MPa	L, m	h , μm	f	I
2.5	4245	143	0.32	$3.37 \cdot 10^{-8}$
2.5	6530	158	0.30	$2.42 \cdot 10^{-8}$
5	1715	139	0.30	$8.10 \cdot 10^{-8}$
5	5442	133	0.26	$2.44 \cdot 10^{-8}$
10	1429	71	0.25	$4.97 \cdot 10^{-8}$
10	2471	169	0.26	$6.83 \cdot 10^{-8}$
20	1768	185	0.22	$1.04 \cdot 10^{-7}$
20	1763	181	0.23	$1.03 \cdot 10-^{-7}$
40	605	170	0.23	$2.81 \cdot 10^{-7}$
40	698	164	0.16	$2.35 \cdot 10^{-7}$

Table 3. Comparative results of tribotechnical tests of samples of ZrO_2 + 3.5 mol% Y_2O_3 crystals after laser treatment with a pulse energy of 4 J and without treatment.

P, MPa	Friction coefficient f		Wear intensity I	
	Without treatment	Laser treatment	Without treatment	Laser treatment
2.5	0.46	0.31	$2.46 \cdot 10^{-9}$	$2.89 \cdot 10^{-8}$
5	0.40	0.28	$2.50 \cdot 10^{-9}$	$5.27 \cdot 10^{-8}$
10	0.36	0.25	$1.02 \cdot 10^{-8}$	$5.9 \cdot 10^{-8}$
20	0.33	0.23	$1.06 \cdot 10^{-8}$	$1.04 \cdot 10^{-7}$
40	0.29	0.19	$4.36 \cdot 10^{-8}$	$2.58 \cdot 10^{-7}$

Laser treatment of PSZ crystals with high-power repetitively pulsed laser radiation in the studied range of pulses with an energy of up to 8 J and a duration of 1.5 - 2.5 ms leads to a structural modification of the surface layer, which is confirmed by previously performed microdiffraction studies. As a result of powerful thermal action, there is a significant change in the phase composition with a high content of the cubic phase (up to 14%). The results of the numerical processing of microdiffraction studies by the EBSD spectroscopy method give reason to believe that the proportion of substances formed as a result of the chemical interaction of the PSZ crystal with the plasma of the near-surface laser flare is no more than 5.8%. A gold-colored coating is formed during laser treatment, the origin of which is a consequence of either the formation of a new chemical compound, possibly zirconium nitride within an unspecified volume of 5.8%, or pure zirconium.

It was established by tribological tests that the near-surface layer formed during laser treatment, in comparison with the tribological properties of an untreated PSZ crystal, has lower values of the friction coefficient and lower wear resistance. Laser modification of the PSZ crystal reduces the coefficient of friction while maintaining the initial high bulk strength of this material. For the coating obtained as a result of laser treatment of the PSZ crystals, during friction without lubrication on У10A steel in the range of normal pressures of 2.5 - 30 MPa, in which the experiments were carried out, the friction coefficient varies from 0.30 to 0.26. When the samples under friction without laser treatment, the friction coefficient varied from 0.45t to 0.28, i.e. the largest decrease in the coefficient of friction was 50%. At the same time, the wear rate increases by 5-10 times. However, it should be noted that in absolute terms, the rate of wear of the PSZ crystal after laser treatment remains quite high and approximately corresponds to the wear resistance of hardened У10A tool steel У10A ($I_{у10A}=5 \cdot 10^{-8}$).

4. Conclusion

The use of laser teatment of PSZ crystals to reduce the friction coefficient is possible with regard to the problems of microtribology, where one of the limiting factors for increasing the reliability of friction units are the antifriction properties of the material in combination with relatively high wear

IOP Conf. Series: Materials Science and Engineering **996** (2020) 012001 doi:10.1088/1757-899X/996/1/012001

resistance, namely, in miniature friction bearings, for example, precision instruments or clock mechanisms.

References

[1] Zammit A *et al* 2019 *Wear* **422–423** 100-7
[2] Hemant K, Rathod J, Nagaraju T, Prashanth K G and Ramamurty U 2019 *Tribology International* **137** pp 94-101
[3] Kümmel D, Linsler D, Schneider R and Schneider J 2020 *Tribology International* **150** 106376
[4] Xu Y *et al.* 2019 *Tribology International* **134** pp 352-64
[5] Piasecki A, Kotkowiak M, Makuch N and Kulka M 2019 *Wear* B **426–427** pp 919-33
[6] Yang K *et al* 2019 Effect of laser remelting on microstructure and wear resistance of plasma sprayed Al_2O_3-40%TiO_2 coating *Wear* A **426-427** pp 314-18
[7] Senouci A, Frene O and Zaidi H 1999 *Wear* **225–229** pp 949-53
[8] Zhao B, KhaderI, Raga R, Konrath G and Kailer A 2019 *Wear* **434–435** 203000
[9] Piasecki A, Kotkowiak M and Kulka M 2017 *Wear* B **376–377** pp 993-1008
[10] Dilawary S A *et al.* 2018 *Wear* **404–405** pp 111-21
[11] Wang J, Du A and Conrad H 2016 *Ceramics International* B **42** 2092-3
[12] Chen Zhao-Chi *et al.* 2020 *Ceramics International* A **46** 17651-8
[13] Kuzminov Yu S, Lomonova E E and Osiko V V 2004 *Refractory materials from a cold crucible* (Moscow: Science) p 369
[14] Milovich F *et al.* 2019 *Materials Chemistry and Physics* **238** 121930
[15] Kuksenova L I, Lapteva V G, Kolmakov A G and Rybakova L M 2001 *Methods of testing for friction and wear* (Moscow: Intermet engineering) p 152

Acknowledgment

This work was carried out with financial support from the Russian Science Foundation (grant 18-13-00397).

IOP Conf. Series: Materials Science and Engineering **996** (2020) 012002 doi:10.1088/1757-899X/996/1/012002

The effect of structural factors on tribological behaviour of copper-zinc gas-dynamic coatings

V E Arkhipov, G V Moskvitin and M S Pugachev

Mechanical Engineering Research Institute of The Russian Academy of Sciences, Moscow, Russian Federation

GVMoskvitin@yandex.ru

Abstract. The paper studies the effect of the structure of copper and brass coatings, obtained by gas-dynamic spattering, on their tribological behaviour. Using durometrical, chemical and X-ray analyses, as well as the wear tests, the authors have determined that the copper coating structure hardness is 102.7 HV, it has pores up to 2 μm in diameter, brass coating consists of 106.5 HV copper, 49.7 HV zinc and 168.7 HV non-metallic compositions (γ – and ε – phases), distributed by mass as 62:7.9:24.2 % correspondingly. The both coatings have presented good results in sliding friction couple tests.

1. Introduction

In the process of coating application by gas-dynamic spattering using low-pressure equipment, the mechanical mixture of pure metals and aluminum oxide is used [1]. A chemical compound (corundum – Al_2O_3) is injected into the mixture of powders to improve the mechanical properties of the coating (adhesion and cohesion). Corundum, striking a part (item), cleans the surface from non-persistent pollutants, oxides, etc., activating it, changing the micro-relief and, therefore, improving the cohesion hardness of the applied metal coating. In the process of interaction of corundum particles with the metal surface and collision between each other, they destroy and partly fix in the coating.

Tribological tests of copper coating applied by gas-dynamic spattering using the mixture of copper and corundum powders, in the media of liquid and consistent grease joined with the counterbody of steel 52100 thermally processed at 58-62 HRC, have revealed that the wear of the applied copper coating was comparable with the wear of steel 52100 after glow-discharge nitriding [2]. In this case the quality of the surface of the counterbody made of steel 52100 thermally processed at 58-62 HRC rises by one class if joined with the coating of copper.

The authors suggest the model of friction surface interaction, according to which the copper coating with the inclusions of fine particles of corundum affects the counterbody as an abrasive tool (emery stone), which results in the improvement of the counterbody surface quality.

The results of tests of the coating based on the mixture of copper, zinc and corundum ("brass") particles, applied on the base coat of 40Cr steel under the conditions of dry friction at the ambient temperature, have shown that its wear is four times lower than that of copper coating [3]. However, the intensity of the counterbody wear is by an order lower if coupled with the "brass" coating. Micro-X-ray spectrometry of the counterbody surface after the test shows the presence of mass transfer of zinc and copper from the coating to the counterbody [4].

The test results are to be assessed as positive, which presupposes the continuation of the study of the possibility of using the coatings, applied by gas-dynamic spattering, in sliding friction units. Tribological behavior of the coatings based on copper and the mixture of copper and zinc particles ("brass") is certainly related to the specific features of formation of the applied metal structure, which may determine the conditions of use and operation of the coatings based on metals and chemical compositions.

The objective of this paper is to study the changes in the coating structure depending on the composition of the spattered mechanical mixture using the particles of copper, zinc, mixture of copper and zinc particles ("brass") and assess the effect of structural factors on tribological behavior of the applied metal coating.

2. Methods and equipment

The coating is spattered on the base coat of steel 5140 using gas dynamic facility DIMET-404 at the rate of interchangeable nozzle displacement in relation to the sample surface 10 $mm \cdot s^{-1}$ and the distance from the face of the nozzle to the surface 10 mm. The air flow temperature is 360° C when spattering a mechanical mixture of particles of copper, zinc and corundum (grade C-01-11), copper and corundum (grade C-01-01), zinc and corundum (grade C-00-11). A mixture of metal particles and a chemical compound is used as received, and the ratio of ingredients by weight is - $Cu:Zn:Al_2O_3$ = 35%:35%: 30%; $Cu:Al_2O_3$ = 55-60%:45-40%; $Zn:Al_2O_3$ = 55-60%:45-40% [1].

The phase composition of the coatings is examined at a Rigaku Ultima IV multifunctional X-ray diffractometer using CuKα-radiation and a parallel beam. X-ray optics in the form of a multilayer parabolic mirror is used to form a parallel beam. Diffraction patterns are taken in symmetric mode, scanning range 2θ from 20 to 120, step 0.05°, detector rate 3° min^{-1}.

The diffraction patterns are analyzed on the basis of the PDXL (Rigaku) software product using the PDF-2 database. The quantitative analysis is performed using the Rietveld method implemented in the PDXL (Rigaku) software package.

The analysis of diffraction lines broadening to determine the parameters of the fine crystal structure (block size, microstrains) was performed using the PDXL (Rigaku) software product by the Rietveld method. The instrumental broadening was recorded by the measurement of the reference - lanthanum hexaboride (LaB_6), that has no physical broadening.

The study of the structure and elemental composition of the coating is performed at FEI Quanta-650 scanning electronic microscope with an EDAX energy-dispersive X-ray spectrum analyzer. The study uses a back-reflected electron detector providing an image with atomic number contrast and visualizing various phases on the surface of the samples. A tungsten cathode served as a source of electrons in the microscope; the accelerating voltage was 25 kV.

The coating metal hardness is measured by Vickers method according to GOST 2999-75 at the load of 245 mN and a holding period of 10 s at the hardness tester SHIMADZU HMV-2. The study is performed from the surface of the coating by two structural components, having a red (copper) and light (zinc) shade.

3. Research results

X-ray phase analysis of coatings based on the mixture of copper and corundum, zinc and corundum, as well as the mixture of particles of copper, zinc and corundum shows that the copper coating contains almost no significant amounts of any compounds and phases.

When a zinc-based coating is applied, aluminum oxide (corundum) in the amount of 5.6% and zinc oxide with the mass fraction of 2.3%, are detected in the metal layer. The analysis of the coating based on the mixture of copper and zinc particles in the coating shows the presence of copper and zinc in the amount of 62.0% and 7.9%, respectively. The amount of corundum in the "brass" coating is almost the same as in the zinc-based coating. Moreover, in the process of coating application, the processes of diffusion of copper into zinc have taken place in the coating with the subsequent formation of intermetallic compounds of γ - and ε - phases, which are formed on the process of cooling of copper and zinc liquid alloy with zinc content of more than 55 atom per cent [5].

The size of substructure of copper and zinc for all coatings exceeds 200 nm; in zinc coating, the oxide size is 242 ± 49 nm. In gamma-phase, the size of the substructure is >200 nm, just like in zinc; in epsilon-phase, it is much less: 82.8 ± 41.5 nm.

The lattice constant of copper (initial cubic face-centered lattice of the constant of 3.608 nm) increases to 3.616 nm after the spattering of copper coating, which is associated with the metal deformation by the particles of corundum. After spattering of a coating based on the mixture of copper and zinc particles, the lattice size still slightly increases (3.620 nm). In the initial composition of a three-component mechanical mixture (Cu, Zn, Al_2O_3), the amount of corundum is significantly lower (30%) than that after spattering of a two-component mechanical mixture (Cu, Al_2O_3), where the amount of corundum is in the range of 40-45% (see the Experiment Procedure section), therefore, any additional effect of corundum on the copper deformation can be excluded.

The copper and zinc hardness do not significantly differ at normal temperature, and, as the metal particles temperature does not exceed 80°C [2] when reaching the base coat surface, the mechanical effect of zinc on copper with the increase of its deformation degree can be omitted. Therefore, the lattice constant increase should be referred to the processes of diffusion and the beginning of generation of solid zinc solution in copper (α is the solid of the solution).

The initial hexagonal close-packed zinc lattice of 2.6649/ 4.9468 nm significantly increases after spattering of zinc coating, especially by one of the parameters – 2.6656/ 4.9494 nm, which is preconditioned by the metal deformation by corundum particles. After the spattering of the coating based on the mixture of copper and zinc particles ("brass"), the size of the lattice increases even greater – 2.6664/4.952 nm, which is preconditioned by the formation of hard copper solution in zinc (η–phase) with the subsequent formation of electronic inter-metal compounds.

The copper coating structure contains a minor amount of pores of the size up to 10 μm, which have formed in the process of mechanical treatment of the applied coating of metal as a result of separation of corundum particles from the matrix, and numerous structural defects of the size not exceeding 2 μm (figure 1).

The authors have analyzed the content of copper, aluminum and oxygen by the path that visually contains no corundum, at the points located on the copper particles (№ 1-3, 8, 9) and at the particles interface (№ 4, 7) (figure 1). The copper content in the spectra № 1-3, 8, 9 is stable enough, ≥98.5 atom per cent, and the presence of aluminum can be described as the "traces". At the interface between the particles the copper content significantly decreases (91-94 atom per cent), and the aluminum and oxygen content rises (spectra 4 and 7). Considering the stoichiometric ratio of aluminum and copper, it is safe to say that at the copper particles interface (spectra № 4, 7) there are the aluminum (corundum) and copper oxides.

From the ratio of the content of components in the points adjoining the interface (spectra № 5, 6) it is possible to judge about the presence of finer particle of aluminum oxide and copper oxide. The size of spectrum initiation zone at micro-X-ray spectrum analysis varies for different elements (it depends on atomic weight, atomic number, density, etc.), it is in the range from 0.2 to 2 μm. The calculations performed show that for copper, the size of spectrum initiation (excitation) zone is ≈2 mm. Considering the structure of the coating (figure 1) and the ratio of the spectra of metals (spectrum 4), it is highly likely, that the aluminum oxide particle size is less than 1 μm.

It should be pointed out that the presence of copper oxide at the interface of copper particles is preconditioned by the use of copper particles which are coated with oxide film at the initial state.

The values of copper coating hardness vary greatly from 72 HV to 122 HV, and the average hardness value is ≈ 102.7 HV.

Considering the results obtained, the following can be noted:

- numerous works in tribology show that the pores play the roles of reservoirs for grease which, in the course of wear of the material and the friction heating and heat expansion of a part (coating) are pressed out from its volume and pass to the friction area, providing the restoration of grease film thus providing the conditions of interface grease and operability of the joined couple [6, 7];

- the surfaces of sliding friction bearings (parts) are normally processed by a tool to 9-10 degree of finish, when the size Rz = 1.6 – 0.8 μm (0.32-0.16 Ra) and, therefore, so fine particles of corundum, even when separated from the matrix (copper), should not affect the wear and the counterbody surface quality deterioration.

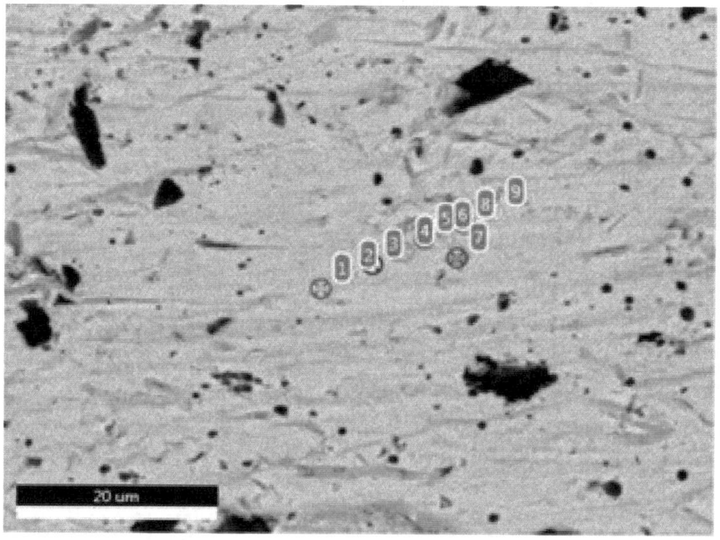

Figure 1. The order of spectra location in the coating based on copper and corundum.

Zinc coating presents more defects (pores), than the copper coating, a number of which contains corundum particles; in average, the size of the structure defects is larger than in copper coating (figure 1, 2). The authors have analyzed the content of zinc, aluminum and oxygen by the pass that visually contains no corundum particles; the spectra in the points located on the particles of zinc (№ 1-3, 6-8) and at the interface between particles (№ 4, 5) (figure 2) have been excited.

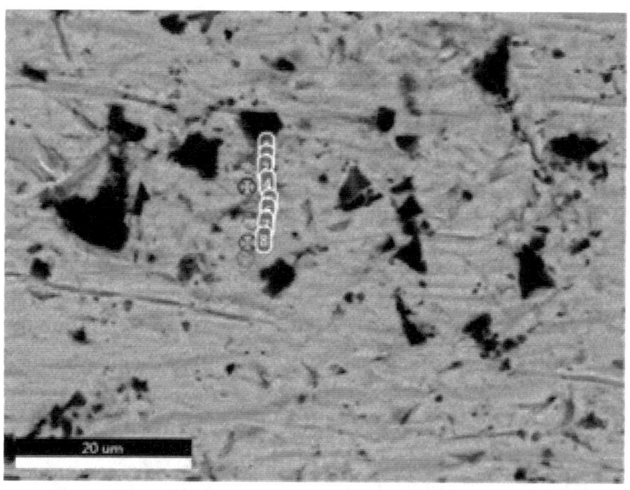

Figure 2. The order of spectra location in the coating based on zinc and corundum.

Zinc content at points № 1-3, 6-8, located on the particles, varies significantly from 55 atom per cent to 97 atom per cent. Considering the change in the content of aluminum and oxygen in the spectra, we can state that corundum particles of different sizes are located at these points. At point 8 it is possible to assume the presence of zinc oxide from the ratio of the content of aluminum, oxygen and zinc. At the interface between zinc particles (spectra 4, 5), there are larger corundum particles than over the body of the particle.

The hardness of zinc coating does not significantly vary, the hardness of the coating is ≈49.7 HV.

In the coating applied using a mixture of copper and zinc particles ("brass"), the pores and corundum particles are visible along the interfaces of copper particles, the size of which does not exceed 5 μm. The corundum inclusions of such size are not observed when analyzing the structure of copper-based coatings (figure 1), the detection of such inclusions during the study of the structure based on a mixture of copper and zinc particles (figure 3) is possible due to the presence of zinc in these zones, which "captures" the corundum particles. X-ray diffraction phase analysis of a coating based on zinc and a mixture of copper and zinc particles showed almost equal corundum content, which indicates the predominant arrangement of corundum particles in zinc particles.

The amount of zinc as well as zinc-based phases in the structure is not so great, only 7.9%, therefore the micro-X-ray spectral analysis of elements distribution (spectra № 1-8) was performed for copper particles, including the interface between them (figure 3).

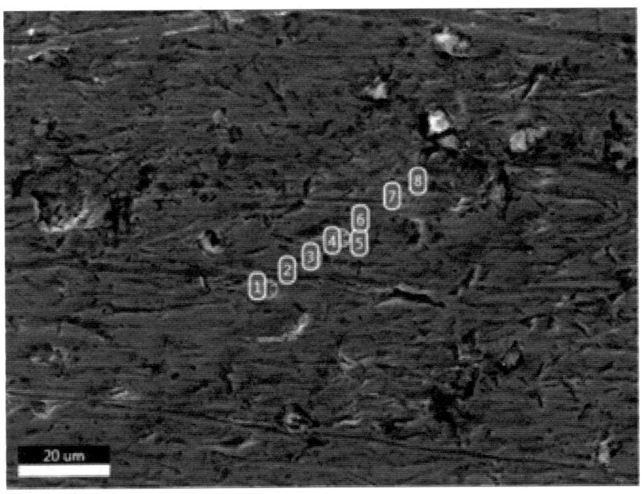

Figure 3. The order of spectra location in the coating based on copper, zinc and corundum.

The spectral analysis of copper particles provides very interesting results: they show the presence of zinc, which is distributed over the surface quite uniformly. Its content ranges from 1.04 atom per cent to 3.52 atom per cent, and a slight deviation in zinc values suggests that the surface of a copper particle is covered with a zinc film, which, in terms of oxygen content, is a zinc oxide. X-ray phase analysis of the coating based on copper and zinc particles did not show the presence of zinc oxide, unlike the coating based on zinc particles. Hence, it is possible to conclude that it is very insignificant, and then the film is of insignificant thickness. The depth of the spectrum initiation zone in micro-X-ray spectral analysis is the same as the spot size 0.2-2 μm. Considering the ratio of the amount of metals (copper and zinc), revealed during the analysis, and the ratio of the spectrum initiation depth to the film thickness, it is possible to calculate that its thickness does not exceed 78 nm. Considering the process of spattering and the subsequent mechanical treatment of the surface for the study, it is possible to say that its appearance can only be associated with the process of zinc mass transfer (displacement) during the treatment of the surface with emery stone and a polishing wheel. In this case, we can note the possibility of mass transfer of zinc when the joining surface affects the coating.

At point 6, which is located at the interface of the particles, we observe the decrease in the copper content and an increase of oxygen and aluminum content, which can be interpreted as the presence of a corundum particle. The amount of aluminum in the other spectra should be considered insignificant, and the increased oxygen content can be attributed to the presence of zinc and copper oxide. The electrochemical potential of zinc is -0.763, and that of copper is +0.552, which, at all other conditions of coating application being equal, preconditions the preferable possibility of zinc oxide generation of coating deposition.

The average hardness of copper particles at significant scatter of values is ≈106.5 HV, which does not differ much from the hardness of copper at spattering of a coating based on a mixture of copper and corundum particles. The hardness of copper does not differ at spattering a mixture of different compositions, whereas the hardness of zinc in the structure of a coating based on a mixture of copper, zinc and corundum particles can be considered equal to the hardness of zinc when spraying a coating based on the mixture of zinc and corundum particles, that is, ≈49.7 HV. The diffusion process taking place during the spattering of the coating results in the formation of intermetallic compounds based on zinc - gamma and epsilon phases with the hardness of ≈168.7 HV.

Studying the structure of coating based on the particles of copper and zinc ("brass"), the authors detected the presence of the structure gradient in terms of composition and hardness, the film of zinc oxide on the copper particles surface and the porousness. Assessing the results obtained, we can point out the following [6, 7]:

- the presence of porousness can also be estimated as a positive factor of "brass" coating operation under the conditions of friction couple grease;
- the materials (alloys) containing both hard inclusions, preventing the basis destruction due to the effect of the abrasive, and plastic ones, which absorb the abrasive, thus improving the conformability of friction surfaces, are widely known in tribology. Therefore, the presence of two phases of various hardness (zinc and electronic compositions) can be assessed as a positive factor under dry friction conditions;
- mass transfer of metals is considered in tribology as a positive factor, as this mechanism provided the cure of the defects that were present or occurred on the surface due to the work of the joined friction couple;
- the presence of zinc oxide film on the surface of the particles may improve the conformability of the joined friction couple.

4. Conclusions

In the process of spattering of a coating with the use of a mechanical mixture of particle of copper and aluminum oxide (corundum), a structure of the hardness ≈102.7 HV is formed. Numerous pores are observed in the structure of the applied metal coating, the typical size of the pores does not exceed 2 μm; the corundum particles of minor size are revealed only over the interfaces of the particles. Such coating can have a good wear resistance in the process of operation of sliding friction surfaces under grease condition due to the presence of grease pockets. The presence of minor (below 1 μm) corundum particles should not affect the wear of the joined couple and the surface quality.

In the process of spattering of a coating with the use of a mechanical mixture of particle of copper, zinc and aluminum oxide (corundum), a structure is formed:

- on the basis of copper – hardness ≈106.5 HV, zinc - ≈49.7 HV, intermetallic compounds (γ - and ε - phases) - ≈168.7 HV, of the weight content of 62.0%, 7.9% and 24.2% correspondingly;
- a low-thickness zinc oxide film is detected on the copper particles which suggests the mass transfer of zinc; a non-significant number of corundum particles of the size below 2 μm is present over the interface of copper particles;
- the "brass" coating of gradient structure may quite efficiently prevent the surface destruction under the effect of hard particles.

The coatings based on copper and the mixture of copper and zinc ("brass") particles need the further study of their tribological properties and characteristics for assessment and specification of conditions of their operation.

References

[1] Dimet. Application of technology and equipment (electronic resource) Available at: http://www.dimet-r.narod.ru/ (accessed 23 December 2019)
[2] Arkhipov V E, Landarski A F, Moskvitin G V and Pugachev M S 2017 *Gas-dynamic coating: structure and properties of coatings* p 240
[3] Sachek B Y, Mezrin A M, Arkhipov V E and Londarskij A F 2018 *Journal of Friction and Wear* **4** pp 299-303
[4] Sachek B Y, Arkhipov V E, Mezrin A M, Muravyeva T I and Shcherbakova O O 2019 *Promising methods of surface treatment of machine parts* p 448
[5] Efremov B N 2016 *BRASSES from phase structure to structure and properties* p 314
[6] Denisova N E, Shorin, V A and Gontero I N 2006 *Tribotechnical materials science and tribology* p 286
[7] Chichinadze AV, Braun E D and Bush N A 2001 *Basic tribology (friction, wear, lubrication)* p 312

IOP Conf. Series: Materials Science and Engineering **996** (2020) 012003 doi:10.1088/1757-899X/996/1/012003

The influence of thermodynamical characteristics of high speed steels on temperature and forces values when turning construction steel

A N Beskopylny, E V Fominov, C G Shuchev and M S Egorov

Don State Technical University, 1, Gagarin sq, 344000 Rostov-on-Don, Russian Federation

fominoff83@mail.ru

Abstract. The article is devoted to the critical problem of studying the dependencies of temperature and cutting forces on the entropy of tool materials (high-speed steels - HSS). The high wear resistance of the cutting tool is closely related to the entropy of tool materials. However, the relationship of entropy as structural sensitive characteristic with temperature and forces during machining processes is poorly investigated. The experimental studies were carried out on the engine lathe with an infinitely variable regulation of spindle rotation frequencies, equipped with a measuring complex, allowing cutting forces and temperature recording at different cutting speeds. It has been established experimentally that at all cutting speeds in the studied range the smallest values of cutting forces and contact temperatures were characteristic for cutting tools made of HSS with high levels of thermal entropy. Thus, the thermal entropy of HSS can be used to the preliminary assessment of relative values of cutting forces and temperatures during the processing of steels which characteristics are strongly connected with wear rates and reliability of cutting tools.

1. Introduction

In modern automated production, a comparatively high proportion of cutting tools, especially small-size tools, is retained by tools, the working part of which is made of HSS. One of the main current trends of improving operating characteristics of cutting tool materials (CTMs) is to create new grades of HSS by improving their chemical composition with new alloying elements. The main channel of energy dissipation in the cutting zone during machining processes is thermal, and quantitative evaluation of the intensity of these processes can be made by temperature values estimation. The temperature values are affected by the load-speed parameters of the cutting process (cutting speed, depth, feed), the most significant of which is the cutting speed. It is the thermal factor, which is significantly dependent on cutting forces, that contributes to the increase in wear rates, and limits the cutting speed value when achieving heat resistance of tool material [1,2]. Currently, the performance properties of HSS tools, including their heat resistance, wear-resistance, and reliability, are improved by chemical-thermal treatment, deposited coatings, laser hardening methods *etc.* [3-5]. The relative increase of performance properties values of modified HSS tools in comparison with tools made of basic HSS grades are estimated mainly on the basis of experimental data. It is important to be able to predict *a priori* some operating and tribological properties of existing and newly developed HSS depending on their chemical composition by investigation of thermodynamic processes in the zone of friction during cutting and receiving analytical dependencies for evaluation of friction and wear characteristics. The study of thermodynamic processes during friction and cutting has enabled the calculation dependence for the wear intensity to be obtained on the basis of the entropy balance equation [6]

$$J = \frac{\left\{\int_0^h (P[S] - \Phi[S])dx\right\}}{v \cdot S} \cdot k, \quad (1)$$

where P[S] – entropy production; Φ[S] – entropy flux; v – sliding (cutting) speed; S – original thermal entropy of the material; k – coefficient, taking into account the share of energy spent on wear.

From (1) it follows that the highest wear resistance values would be obtained for CTMs with higher values of the original thermal entropy S. Thus, in the development of new HSS grades preference should be given to compositions characteristic by high levels of thermal entropy, which calculation is not difficult for heterogeneous structures of known chemical composition under the additive rule [7,8].

High-entropy CTMs also characterized by low values of absolute (relative) thermo-EMF and high surface dynamic micro hardness demonstrate better tribological characteristics and higher wear resistance in comparison with low-entropy materials [6,9-11]. They also tend to form highly effective shielding dissipative structures and show low values of friction coefficient for some of the materials being processed, including structural steels [11,12].

The aim of this work is to study the dependencies of cutting forces and temperature values in the cutting zone on cutting speed during processing work-pieces made of structural steel 45 considering the thermal entropy value of HSS as an integrated characteristic of CTM chemical composition. This study is part of a set of research into the cutting, operational and tribological properties of high-entropy CTM.

2. Materials and methods

The research was carried out on a lathe without using of cutting fluids. The exact values of the cutting speeds V were obtained using the frequency converter Mitsubishi. The spindle rotation frequencies n were controlled by the MEGEON 18005 laser tachometer with an accuracy of 0.5 rpm. To measure and record cutting force components Pz, Py, Px and cutting temperature T the STD.201-1 (Armenia) measuring system, equipped with the National Instruments interface unit obtaining the signal sampling frequency of 0.78 kHz, was used. The turning work-pieces of 45 structural steel (analog of steel grades 1044, 1045 – USA, 1.0503 – BRD and S45C – Japan) was made using specially made whole tools of 7 grades of HSS with different values of thermal entropy S. The grade's denomination, chemical composition and thermal entropy S of the HSS specimens are represented in the table 1.

Table 1. Studied HSS grades.

№	Grade	Chemical composition	Thermal entropy S, J·mol^{-1}·K^{-1}
1	R6M3	W(6%)+Mo(3%)+Fe	26.86
2	R6M5	W(6%)+Mo(5%)+Fe	27.26
3	R6M4F4	W(6%)+Mo(4%)+V(4%)+Fe	27.46
4	R8M3F4	W(8%)+Mo(3%)+V(4%)+Fe	27.53
5	R18	W(18%)+Fe	28.04
6	EP658	W(6%)+V(2%)+Co(8%)+Mo(5%)+Fe	30.60
7	EP657	W(12%)+V(2%)+Co(8%)+Mo(3%)+Fe	30.78

The linear dimensions and the cutting part geometry of the HSS specimens are presented on figure 1. Cutting depth t=0.5 mm, feed s=0.15 mm·rev^{-1}.

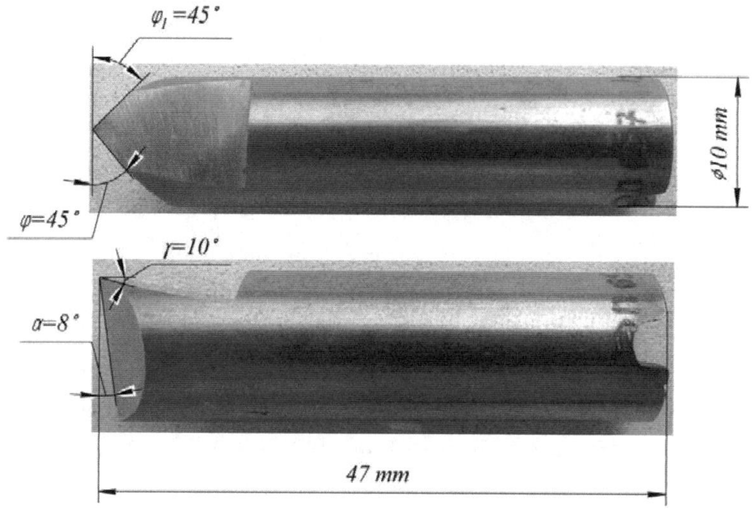

Figure 1. The linear dimensions and the cutting part geometry of the HSS specimens.

The cutting temperature T was measured using the natural thermocouple method. Preliminary taring was made by heating each of the seven "HSS - Steel 45" thermocouples and corrective coefficients were obtained to determine the temperature values T in the cutting zone based on the relative thermo-EMF values recorded by the measuring complex. The statistical processing of the data was carried out by an algorithm composed of standard MathCAD functions for analysis of variance (ANOVA) using the methodology [13].

3. Results and discussion

Comparison of cutting forces for HSS of different grades was made by the average values of Pz_{avg}, Py_{avg}, Px_{avg} without taking into account the initial and final cutting stages (figure 2).

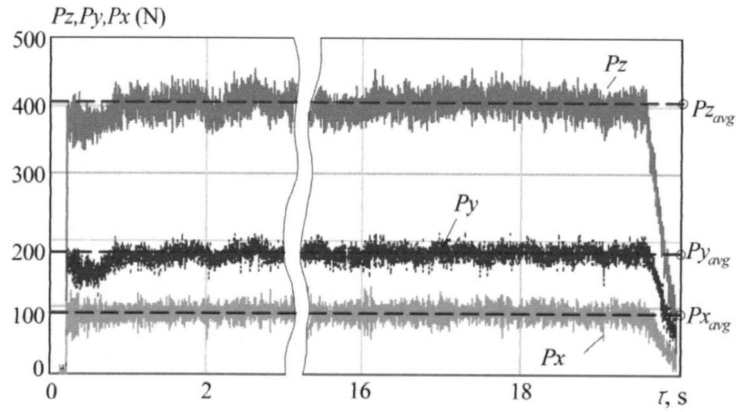

Figure 2. Cutting forces components Pz, Py, Px variations in time τ when turning by R18 tool (S=28.04 $J \cdot mol^{-1} \cdot K^{-1}$), cutting speed V=50.3 m·min^{-1}.

Figure 3 (a), (b), (c), (d) shows the dependencies of the cutting forces and temperature in the cutting zone on the cutting speed. It is seen that the lowest averages of Pz, Py, Px were recorded in the case of high-entropy HSS tools. Moreover, with the increase of the cutting speed value component

Pz_{avg} for high entropy HSS tools also increases, while for low entropy HSS tools there is a tendency to decreasing.

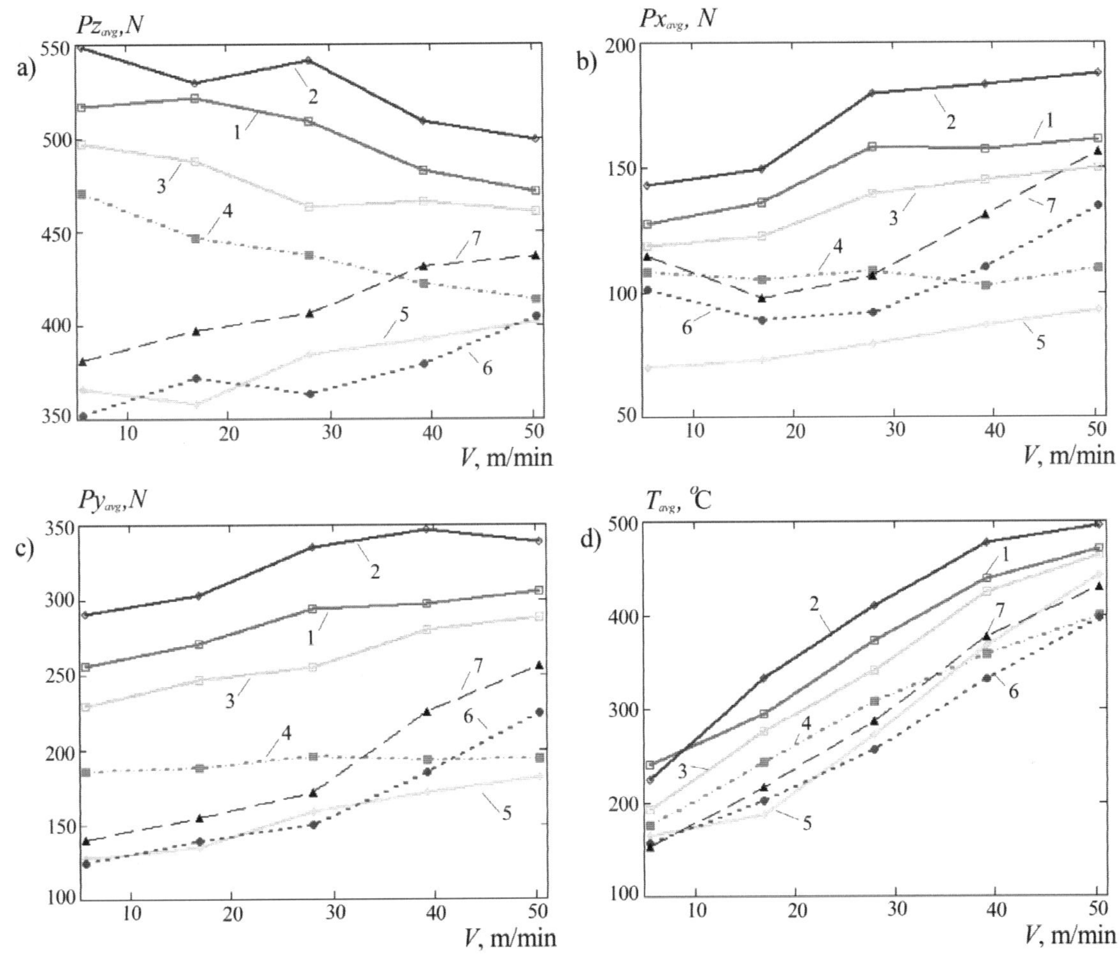

Figure 3. The dependences of Pz_{avg} (a), Py_{avg} (b), Px_{avg} (c) and contact zone temperature T_{avg} (d) on cutting speed V when turning steel 45 by HSS tools of different grades: 1 – R6M3; 2 – R6M5; 3 – R6M4F4; 4 – R8M3F4; 5 – R18; 6 – EP658; 7 – EP657.

The cutting force components Py_{avg} and Px_{avg} for all investigated HSS grades tend to grow with increase of cutting speed V. Thc high-entropy HSS are also characterized by lower temperature values T in the contact zone at all cutting speeds V of studied range (figure 3, (d)).

The presented data are well consistent with results of previous investigations of tribological properties of HSS specimens of the same grades under conditions of friction on structural steels without lubricant. Thus, as a result of tribological tests, when the levels of sliding speed and normal load were close to those characteristic to machining processes, lower friction coefficients, temperatures and wear intensity rates were recorded for high-entropy HSS specimens [6,10]. The same dependence of tribological characteristics on the HSS entropy was observed as a result of tests performed on tribometre which implemented the scheme of friction "pin on disc" at comparatively low pressure values. In this case, formation of dissipative shielding structures of considerable thickness, contributing to the separation of rubbing surfaces and acting as dry lubricant has been recorded for the specimens (pins) made of high-entropy HSS grades [12].

4. Conclusions

As a result of the study it has been established experimentally that at all cutting speeds in the studied range the smallest values of cutting forces and contact temperatures were characteristic for cutting tools made of HSS with high levels of thermal entropy. Thus, when choosing CTM for structural steels machining, and in the development of new HSS grades preference should be given to compositions characteristic by high levels of thermal entropy as obtaining better tribological characteristics and higher wear resistance together with potentially higher cutting speed and productivity values in comparison with low-entropy materials.

References

[1] Grzesik W 2017 *Advanced Machining Processes of Metallic Materials: Theory, Modelling, and Applications* (Elsevier)
[2] Ryzhkin A A *et al* 2007 *Material cutting* (Rostov-on-Don: DGTU Press)
[3] Baek G Y *et al* 2020 Effect of Post-Heat Treatment on the AISI M4 Layer Deposited by Directed Energy Deposition *Metals* **10** 703
[4] Soffritti C *et al* 2020 Influence of Vacuum Heat Treatments on Microstructure and Mechanical Properties of M35 High Speed Steel *Metals* **10** 643
[5] Liu B *et al* 2019 Effect of Tempering Conditions on Secondary Hardening of Carbides and Retained Austenite in Spray-Formed M42 High-Speed Steel *Materials* **12** 3714
[6] Ryzhkin A A, Burlakova V E *et al* 2016 Determination of the efficiency of high-entropy cutting tool materials *J. Frict. Wear* **37** pp 47–54
[7] Karapetyants M H 1975 *Chemical thermodynamic* (Moscow: Himiya Press)
[8] Honig J 2013 *Thermodynamics 4th Ed* (Elsevier)
[9] Ryzhkin A A, Burlakova V E and Novikova A A 2018 Wear and performance of hard alloys *Russ. Engin. Res.* **38** pp 438–41
[10] Ryzhkin A A, Fominoff E V and Torop Y A 2017 Estimation of triboelectrical properties of high-speed steels *Vestnik DGTU* **2** pp 20–29
[11] Beskopylny A N, Fominov E V, Shuchev C G and Ryzhkin A A 2019 Tribological characteristics of experimental hard alloy grades with modified cobalt binder under conditions of dry friction on titanium-aluminum alloy VT-3 *Proc. Int. Conf. SerbiaTrib* pp 49–58
[12] Ryzhkin A A, Fominoff E V and Shuchev C G 2018 Study on the tribological characteristics of high entropy high speed steels in conditions of dry friction on structural steel *Proc. Int. Conf. ICIE* 1819–27
[13] Montgomery D C 2012 *Design and Analysis of Experiments* 8th Ed (John Wiley & Sons Inc.)

IOP Conf. Series: Materials Science and Engineering **996** (2020) 012004 doi:10.1088/1757-899X/996/1/012004

Influence of laser quenching modes on tribological characteristics of 40Cr steel

V P Biryukov[1], S N Sokolov[2], A P Savin[1] and A N Prince[1]

[1] Mechanical Engineering Research Institute of The Russian Academy of Sciences, 4 Maly Kharitonyevsky Pereulok, Moscow, 101000, Russia
[2] LLC RPA "INZHEKT", 3a Elmashevskaya, Saratov, 410033, Russia

Laser-52@yandex.ru

Abstract. This paper presents the results of metallographic and tribological studies of 40Cr steel samples after laser quenching with a rectangular beam. Using a full factorial experiment (FFE), the regularities of changes in the geometric parameters of the laser thermal hardening zones from the processing modes are obtained. It is shown that the jamming load is 1.5-2.1 times higher, and the coefficient of friction is 2-4 times lower than that of normalized 40Cr steel.

1. Introduction

The application of laser hardening technologies in modern mechanical engineering is an urgent task. Laser hardening of 50-70% of the friction surface area can increase the durability of friction units in power and transport engineering by 2-3 times. Samples of steel containing 1.0% carbon and 1.5% chromium with a size of 11 × 24 × 100 mm were used for laser hardening with a Laserline diode laser, a rectangular spot with a size of 3 × 8 mm, at a radiation power of 2000 W, and a beam travel speed of 20 mm·s^{-1} [1]. The first batch of samples was processed by laser radiation, without preliminary heat treatment (WHT). The second batch was heated to 1000 °C and released at 650 °C (HTQ). The third batch was quenched at a temperature of 500 °C and released at a temperature of 170 °C (QHT). The fourth batch was quenched twice at 850 °C and released at 170 °C (DQHT). The depth of the laser-hardened layer of WHT, HTQ, QHT, and DQHT samples was 462.8 microns, 480.9 microns, 669.5 microns, and 663.2 microns, respectively. The depth of the laser quenching zone of HTQ and QHT samples with preheating up to 160 °C was 578.5 microns and 764.5 microns, respectively. For QHT and DNTO samples, the maximum hardness of the hardened layer 1000 HV was at a depth of 0.5 mm. For WHT and DNTO samples, the maximum hardness within the hardened layer was about 900 HV, at a depth of about 0.3 mm.

In order to investigate the impact of the surface geometries on the characteristics of the heat-affected and hardened zones, laser hardening experiments were conducted at steel 38MnVS6 in the present work using a beam an IPG fiber laser with a power of 2 kW and a beam diameter of 9 mm and speed 10 mm·s^{-1} [2]. For comparison, two types of samples are made flat, and with a groove 6.9 mm wide with a radius of 4 mm and a depth of 2 mm. The depth of the laser hardening zones is determined from optical micrographs and hardness measurements. It was 890 ± 6.5 microns and 897 ± 4 microns for the flat sample, and 719 ± 75 microns and 706 ± 17 microns for the shaped sample, respectively. The maximum measured hardness values were 942 $HV_{0.05}$ (flat surface) and 943 $HV_{0.05}$ (curved surface).

Samples of 44MnSiVS6 steel in the form of disks with a thickness of 10 mm and a diameter of 110 mm with a surface roughness Ra 0.5-1 microns were strengthened using a fiber laser system from IPG Photonics [3]. The maximum laser power was 3000 W, and the beam travel speed was 100 mm$\cdot$s^{-1}. The treatment was performed with three types of laser spot. Round shape with a diameter of 4 mm, with a Gaussian distribution of radiation energy. A 4×4 mm rectangular spot with uniform energy distribution and a 4×4 mm rectangular spot with 16 radiation points, with a total power loss of 37.9% of the round spot. The depth, width, and microhardness of the steel hardening zones were 0.4×1.6 mm, 5800-8000 MPa, 0.3×1.4 mm, 4200-7300 MPa, 0.2×1.1 mm, and 3800-7200 MPa for round, rectangular, and spot spots, respectively. Treatment with a round laser beam resulted in residual compression stresses of up to 506 MPa in the quenching zone to a depth of 400 microns.

Laser surface treatment was carried out on samples of powdered carbon steel containing 1% carbon PA - FeGr with a porosity of 4%, 8% and 10%, as well as for comparison of U10 steel [4]. Powder steel samples were obtained by powder metallurgy from iron powder and colloidal graphite. The powders were mixed in a mixer with an offset axis of rotation, pressed in a mold at a pressure of 400 MPa, then annealed in a vacuum furnace at a temperature of 900°C, re-compacted in a mold at a pressure of 500 or 550 MPa, and finally sintered in a vacuum at a temperature of 1150°C. Laser surface treatment was performed on samples in the form of parallelepipeds 25×25×200 mm in size using an industrial additive machine Optomec LENS 850-R with a 1 kW fiber laser in argon. The diameter of the spot of the laser beam during laser processing was assumed to be 1.5, 2, 2.5 and 4 mm. The laser radiation power was 251-502 W, and the beam travel speed was 8-12 mm$\cdot$s^{-1}. In contrast to U10 steel, microstructure analysis in powder steels allowed us to divide the hardening region into 4 zones: I - melting with reduced porosity, II - melting with initial porosity, III-solidification, and IV-base. Structure in zones I and II, martensitic microhardness is high, up to 1000 $HV_{0.05}$. Porosity does not affect the microhardness of the surface layer, but it affects the depth of hardening, which is greater with a lower porosity, and for a sample of U10 steel, the depth of hardening corresponds to the depth of the heat-affected zone.

The purpose of this work is to determine the effect of laser treatment modes with a rectangular spot on changes in the jamming load from the sliding speed, friction coefficient, and wear intensity of 40Cr steel.

2. Materials and equipment

Laser hardening was performed on an automated technological complex equipped with a 5.9 kW diode laser on 40Cr steel samples with dimensions of 12×30×60 mm. The effect of laser radiation power on the depth (H), width (B), and microhardness HV of the quenching paths was determined in experiments. The variable parameters were the radiation power P = 3000 W and 3640 W and the speed of beam movement V 7.6 mm s^{-1} and 9.8 mm s^{-1}, the deviation of the laser beam from the focusing plane was 19.5 mm. The laser beam in the focal plane had dimensions of 16×4.6 mm. For metallographic studies, a digital microscope Dino-Lite Pro AM423ML, a metallographic microscope, and a microhardness meter PMT-3M with an indenter load of 0.98 N were used.

Tribological characteristics were determined by testing flat laser-hardened samples in a friction pair with ring samples of 95Cr18 steel (59-62 HRC). The pressure and sliding speed changed stepwise in the range of 0.5-3.0 m$\cdot$s^{-1} and 1-5 MPa, respectively. TP22C oil was used to lubricate the friction pair.

3. Results of experiments and calculations

Figure 1 shows a micro-section of a heat-strengthened 40Cr steel laser track. The depth of the quenching zone varied in the range of 0.92-1.41 mm, with a zone width of 18.1-20.91 mm. The shape of the laser spot in the form of a rectangle allows you to get almost uniform depth of the quenching zone. The microhardness of the quenching zones was 5780-7440 MPa.

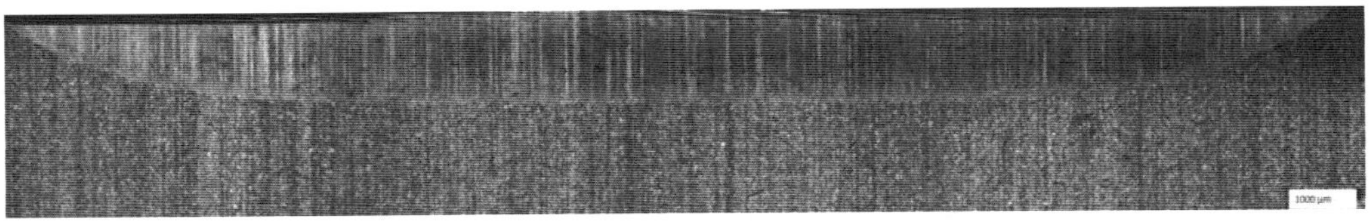

Figure 1. The micro-sections of the zones of hardening of steel 40Cr.

Using mathematical planning of the experiment, the equations [5] are obtained.
Depth of the quenching zone:

$$H = 1.17725 + 0.01625\left(\frac{P_i - 3320}{120}\right) - 0.24675\left(\frac{V_i - 8.75}{1.25}\right), \quad (1)$$

Width of the quenching zone:

$$B = 19.521 + 0.9885\left(\frac{P_i - 3320}{120}\right) - 0.518\left(\frac{V_i - 8.75}{1.25}\right) \quad (2)$$

For figure 2 shows the dependences of the depth (H) and width (B) of the laser-hardened layers on the processing modes.

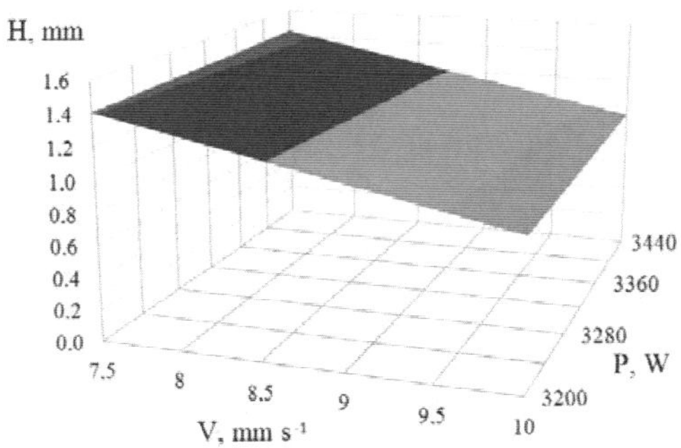

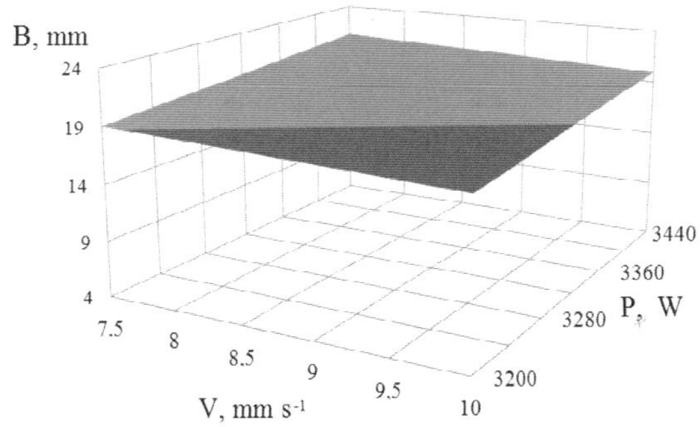

(a) (b)

Figure 2. Dependence of the depth (a) and width (b) of the laser hardening zones of 40Cr steel.

As the beam power increases and the processing speed decreases, the depth and width of the quenching zones increases. The regression equations were used for calculations and compared with the results of the experiment. The calculated values differ from the actual values of the depth and width of the quenching zones by no more than 3.56%.

Figure 3 shows graphs of the dependence of the jamming load on the sliding speed. Laser hardening increases the jamming load in comparison with the initial samples at a sliding speed of 1 $m \cdot s^{-1}$ by 1.6 – 2.1 times, depending on the microhardness of the quenching zones. At a pressure of 1.1 MPa, laser-hardened surfaces are susceptible to jamming at speeds of 2.6 – 3.1 $m \cdot s^{-1}$, while untreated samples are inoperable at speeds of 1.45 $m \cdot s^{-1}$.

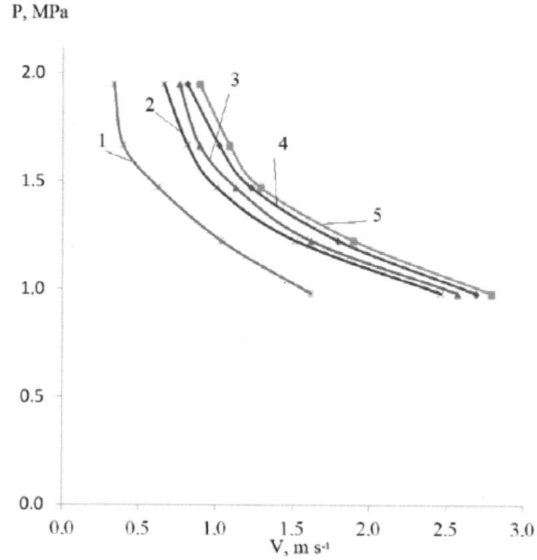

Figure 3. Dependence of the jamming load on the sliding speed: 1-normalized steel, 1800-2040 MPa, 2- 5780-6740 MPa, 3- 6230-7130 MPa, 4- 6350-7210 MPa, 5- 6640-7840 MPa.

One of the most important characteristics of a friction pair is the coefficient of friction. Figure 4 shows the regularity of changes in the coefficients of friction from pressure. As the microhardness of the laser-hardened layer increases and the pressure increases, the coefficient of friction decreases. At a pressure of 3 MPa, the coefficient of friction is 3 to 4 times lower for hardened samples compared to the original steel. With a further increase in pressure to 5 MPa, the value of the coefficient of friction remains 2-3 times less than in normalized steel.

An equally important characteristic of the friction pair is the wear rate. Table 1 shows the results of tests for the wear rate of samples. As the microhardness of the samples increases, the wear rate decreases. Wear resistance, the reverse value of the wear intensity increases 2-2.6 times compared to the base material.

4. Discussion

The results obtained clearly show the possibility of increasing the wear resistance and friction surfaces of machine parts and aggregates during laser quenching. In our country, the most developed technologies are strengthening pulsed solid-state, gas, and fiber laser installations. The creation of a laser complex with a powerful 6 kW diode laser will expand the range of hardened parts.

One of the unsolved problems is laser hardening of gears with a module of 6-10 mm with a depth of the hardened layer of 1.2-1.45 mm in one pass. Laser hardening of the tooth surface with the imposition of tracks can lead to jamming of the engagement. This is due to the fact that when applying laser tracks, tempering zones with a width of 0.5 – 1.7 mm appear, which are the centers of setting of friction pairs, even at small specific loads. The hardening width of 15-21 mm is sufficient to increase the wear resistance by 3-5.5 times of tillage tools, cultivator paws, disc harrows and others.

The introduction of a new laser hardening technology will increase the durability of parts in the energy, aerospace and transport engineering industries.

The paper presents a narrow range of processing power and speed. Further research is needed to expand the processing modes and range of steel grades. Tribological tests will reveal the possibilities of new laser strengthening technologies.

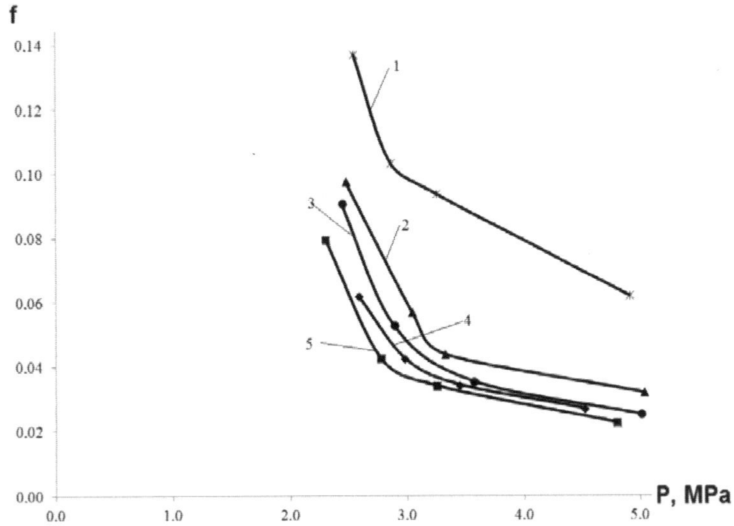

Figure 4. Dependence of the coefficient of friction on pressure: 1-normalized steel 1800-2040 MPa, 2-5780-6740 MPa, 3-6230-7130 MPa, 4-6350-7210 MPa, 5-6640-7840 MPa.

Table1. The wear rate of the samples.

		Intensity wear	
No. sample	Microhardness, MPa	$I_l \cdot 10^{-9}$	$I_{l\,m} \cdot 10^{-9}$
1.1	1820-1970	0.983	
1.2	1800-2020	0.969	1.024
1.3	1860-1990	1.120	
2.1	5780-6650	0.536	
2.2	5820-6710	0.525	0.519
2.3	5690-6740	0.496	
3.1	6320-7020	0.523	
3.2	6440-7067	0.489	0.496
3.3	6350-7130	0.478	
4.1	6350-7120	0.460	
4.2	6430-7090	0.475	0.459
4.3	6390-7210	0.443	
5.1	6640-7380	0.410	
5.2	6720-7690	0.397	0.394
5.3	6680-7840	0.376	

5. Conclusion

The technology of laser hardening of 40Cr steel with a layer depth of 0.9 – 1.4 mm and a quenching zone width of 18 – 20 mm with a rectangular beam has been developed. The scuff resistance of laser hardening zones is 1.5-2 times higher, and the coefficient of friction is 2-4 times lower compared to normalized 40Cr steel.

References

[1] Li Z, Tong B, Zhang Q, Yao J, Kovalenko V and Li Z 2020 *J. Mat. Sci. Eng.* **788**

[2] Joerg V, Sandra D H, Fischer A and Niendorf T 2020 *J. Adv. Manuf. Technol.* **107** pp 4255–60
[3] Dewi H S, Fischer A, Volpp J, Niendorf T and Kaplan F H 2020 *Opt. Laser Tech.* **127**
[4] Oglezneva S, Morozov E and Usynin E 2019 *MATEC Web of Conferences* **298**
[5] Evdokimov Yu A, Kolesnikov V I and Teterin A I 1980 *Planning and analysis of experiments in solving problems of friction and wear* (Moscow: Nauka)

IOP Conf. Series: Materials Science and Engineering **996** (2020) 012005 doi:10.1088/1757-899X/996/1/012005

Determination influence of graphite grade on tribological properties of friction material based on tin bronze when using "A" and ATF-III grade oils

V P Biryukov[1], A F Ilyushchenko[2], A V Leshok[2] and A N Rogovoy[2]

[1] Mechanical Engineering Research Institute of The Russian Academy of Sciences, 4 Maly Kharitonyevsky Pereulok, Moscow, 101000, Russia
[2] State Scientific Powder Metallurgy Institute, 41 Platonova str, Minsk, 220005, Belarus

Laser-52@yandex.ru

Abstract. The paper presents the results of metallographic and tribological studies of tin bronze samples with additives of elemental graphite (GE) and pencil graphite (GP) with an average size of 10 μm in an amount of 10% vol. The wear rate of bronze samples with graphite GI is higher than with GP by 2 times when tested with mineral oil and 1.1 times when tested with synthetic oil. The jamming load is 1.5-2 times higher than bronze with GP graphite compared to bronze without graphite.

1. Introduction

The matrix material used in this work was a commercially pure copper powder with an average particle size of ~35 μm, produced by CDH, India. Graphite powder with an average particle size of ~10 μm was used as a reinforcing material. A copper matrix composite reinforced with graphite powder was produced by powder metallurgy with three different volume fractions of graphite (1, 2, and 3%) [1]. The wear test samples were manufactured in accordance with ASTM G 99-04 standards, and their dimensions were 10 × 20 mm. The experiments were performed at variable loads of 2.5, 5, 7.5, 10, and 12.5 N, a constant speed of 1.178 ms^{-1} (150 min^{-1}), and a duration of 15 minutes for each test. The wear resistance of manufactured composite increases with percentage of graphite particles, which acts as a solid lubricant. At 3% graphite, the wear resistance was improved by about 88%, compared to the copper sample without graphite additives.

Samples from the Cu-Al-Gr composite with dimensions of 17 × 15 mm, containing electrolytic copper powder with a particle size of 50-80 μm, graphite-200-300 μm, and aluminum powder - 70-100 μm [2]. They were manufactured by powder metallurgy at a pressure of 1.77 MPa and a temperature of 820 °C. Tribological tests were carried out according to the "disk-finger" scheme, at a speed from 200 min^{-1} (3.14 ms^{-1}) to 3000 min (47.1 ms^{-1}), and a pressure of 0.5 MPa. A rotating disk with a diameter of 300 mm made of H13 steel (a disk material for high-speed trains) was used as a counterbody. The graphite content was changed from 3 to 18%. The minimum coefficient of friction and wear intensity were obtained with a graphite content of 12%.

Graphite 6, 8, 10, 12 and 15% were added to the charge of the Cu-Sn-Fe-SiO2-MoS_2 friction material. Samples with a graphite content of 10% have a lower strength, but there is a significant increase in their wear resistance [3]. As the rotation speed increases from 200 to 400 min^{-1}, the amount of graphite on the friction surfaces increases and the wear pattern changes.

The purpose of this work is to establish the influence of graphite grade and lubricant on changes in the jamming load from the sliding speed, wear intensity, and coefficient of friction.

2. Materials and equipment

Tin bronze (12% by volume of tin) obtained by mixing copper powders of the PMS-1 brand and TIN of the PO -1 brand with an average particle size of 80 and 10 μm, respectively, was used as the base of the powder friction material. The base was added graphite element grade GE-1 (GOST 7478-75) with an average particle size of 80 μm and graphite pencil GP-1 (GOST 17022-81) with an average size of 10 μm in an amount of 10% vol.

Blanks of three samples with a diameter of 35 mm and a height of 25 mm for tribological tests were made as follows: the initial powders were mixed in a "drunk barrel" type mixer for 2 hours, samples of a given shape were pressed at a pressure of 200 MPa, then sintered in a protective-reducing atmosphere of endogas at a temperature of 840 °C for 240 minutes.

Wear and tear resistance tests were performed on the MTU - 01 friction machine according to the scheme plane (sample size 12 × 20 × 70 mm, steel 65Mn) - ring (diameter 34 mm, height 22 mm, composite powder friction material) at a load of 0.5-3 MPa and a sliding speed of 0.5-4 ms^{-1}. Two types of oil were used as a lubricant: 1-hydraulic mineral oil of the "A" brand (TU 38.1011282-89) for operation in torque converters and automatic transmissions of cars from -40°C to ambient temperatures; 2-transmission oil for automatic transmissions ATF-III (Sintoil). The lubricant was applied to the friction zone at a rate of 1 drop per second. The wear rate was determined by the expression [4]:

$$J = \frac{\Delta h}{2\pi r n} \tag{1}$$

where, Δh - is the average linear wear, over n - test cycles, r - is the average radius of the sleeve sample made of composite sintered material, and n is the total number of revolutions of the mobile sample.

$$\Delta h = h - h_1 = \tan^{-1}\left(\frac{\alpha}{2}\right)(b_0 - b_1) \tag{2}$$

where, h – is the initial depth of dimple (μm), h_1 – is the depth of the worn hole, (μm), b_0 – initial width of the hole (mm), b_1 – width of the worn hole (μm), α – the angle between the surfaces of the walls of the wells. The load on the indenter during scratching was 4.9 N on the PMT-3 device. The speed of movement of the diamond indenter was 10 mm·s^{-1} [5].

3. Results of experiments

Figures 1 and 2 show fragments of samples after friction and the formation of a bully. The friction surface of a sample of composite material with GE graphite has deeper wear grooves towards sliding, and observed pores size 50-200 μm in places where accumulates graphite (figure 2 (a)). Pore formation is associated with the transfer of graphite to the steel surface during friction. The sample with GP graphite has, the pore size smaller 50 μm and the friction surface has less deep risks towards sliding direction (figure 2 (b)).

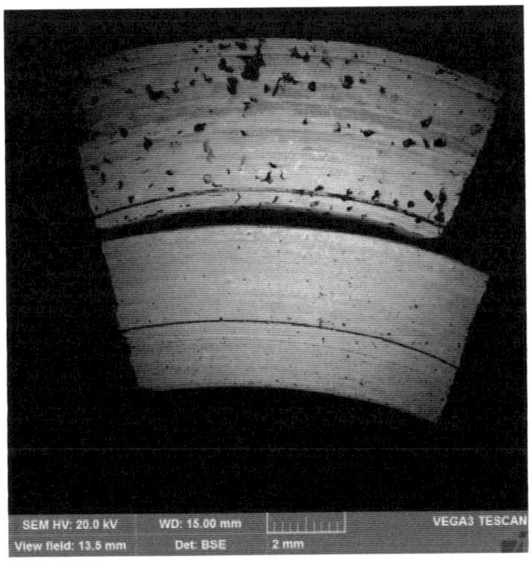

Figure 1. Type of friction surface of samples after tribological tests: upper sample - GE graphite 10% vol; lower-GP graphite 10% vol.

(a)

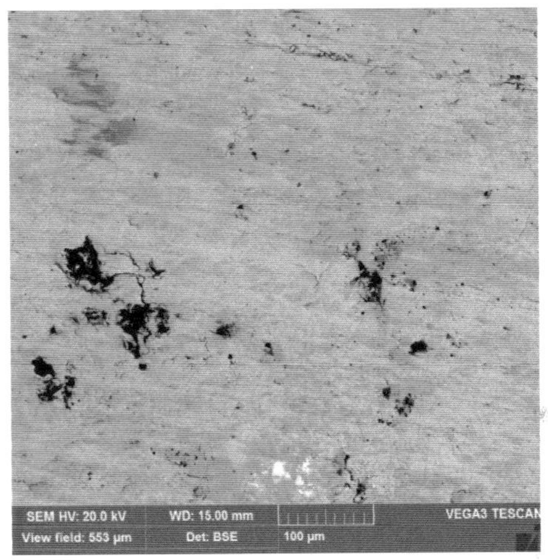

(b)

Figure 2. Fragments of friction surfaces of samples after tribological tests: a-GE graphite, b- GP GP graphite.

Figure 3 shows the zones for determining the elemental composition on samples of bronze and counter-samples of 65Mn steel. On the friction surface of the graphite GE bronze sample, dark zones with an increased carbon content of 6.72 – 7.11% (spectra 1 and 2) containing graphite are observed, and on the counter - sample (figure 3 (c)) in the spectra 13,14 and 18,19, an increased content of copper in the range of 22-56% and graphite in the spectrum 20.

By friction, copper and graphite were transferred on the steel surface. Contain copper in the sample of bronze with graphite GP is in spectra 6-11 (figure 3 (b)) has similar values with similar spectra (figure 3 (a)). In spectrum 12, observed increase content carbon. That indirectly indicates the presence of GP graphite in this region. The transfer of iron to the friction surface is insignificant and amounts to 0.2-0.38% in the spectra 6-8, on the counter – sample of steel 65Mn (figure 3 (d)) in the spectra 31 – 33, the carbon content is in the range of 7 – 17%, and copper 71-77%.

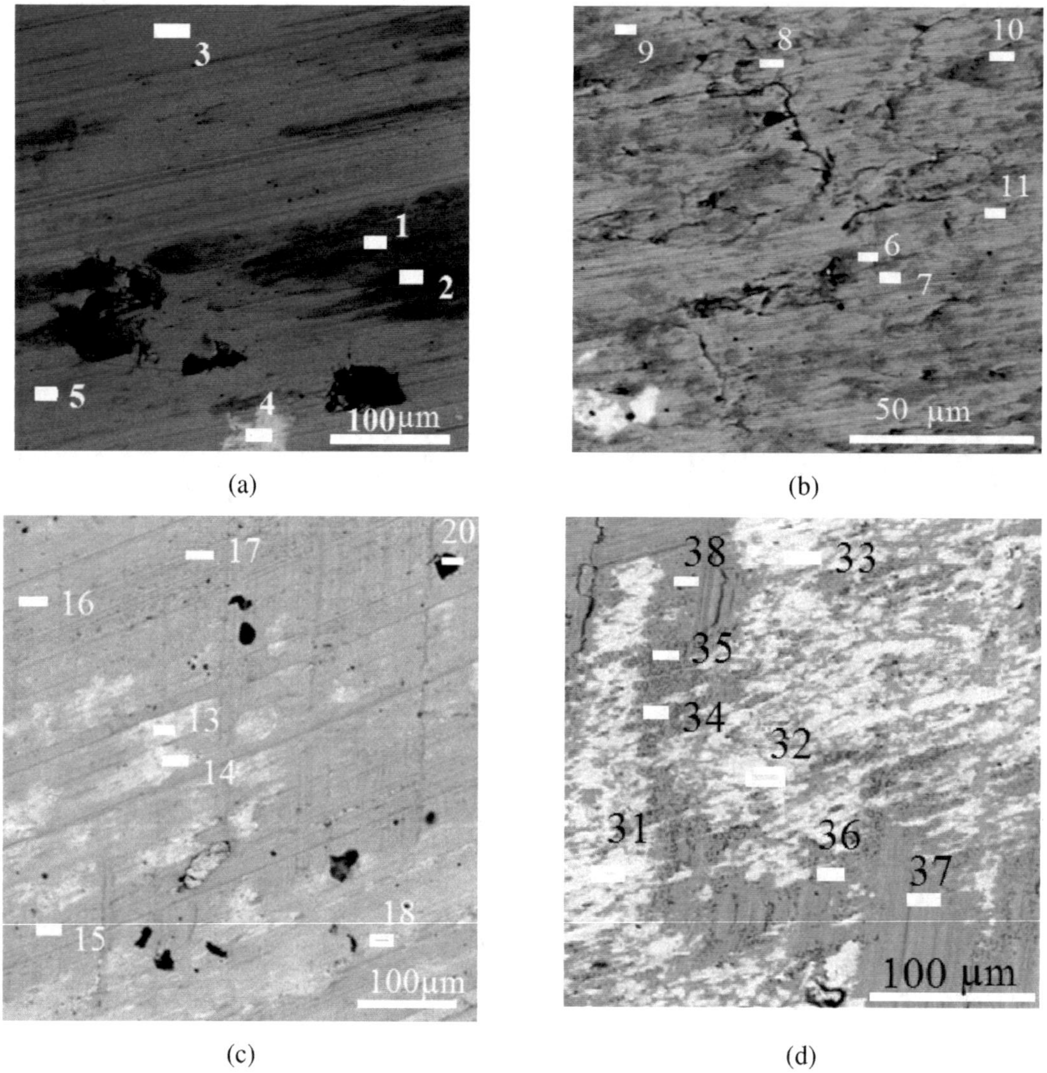

Figure 3. Areas for determining the elemental composition of samples of bronze-graphite material with 12% tin and counter-samples of steel 65Mn: a-bronze + graphite GE, b-bronze + graphite GP, c and d-steel 65Mn.

The results of tests for the wear rate of bronze-graphite formations containing 12% tin and 10% graphite of the GE and GP grades paired with counter-samples made of 65Mn steel are shown in table 3. The Tests were performed at a sliding speed of 0.59 ms^{-1} and a pressure in the friction pair equal to 1 MPa.

The wear rate of bronze samples with GE is 2 times higher than with ha when tested with mineral oil and 1.1 times higher when tested with synthetic oil. The friction coefficients were 0.068 and 0.067 and 0.070 and 0.058 with mineral and synthetic oils, respectively. The estimation of the jamming load from the sliding speed and the type of lubricant is shown in figure 4.

Table 1. Results of wear intensity tests with mineral oil grade "A" / with synthetic oil ATF-III.

№ samples	Sample material	Steel wear rate $I_1 \cdot 10^{-9}$	Wear rate sample $I_2 \cdot 10^{-9}$
1	Bronze 12% Sn, porosity 20%, 10% vol. GE	0.7565 / 0.5578	0.9555 / 1.2333
2	Bronze 12% Sn, porosity 20%, 10% vol. GP	0.7009 / 0.6677	0.5068 / 1.1608

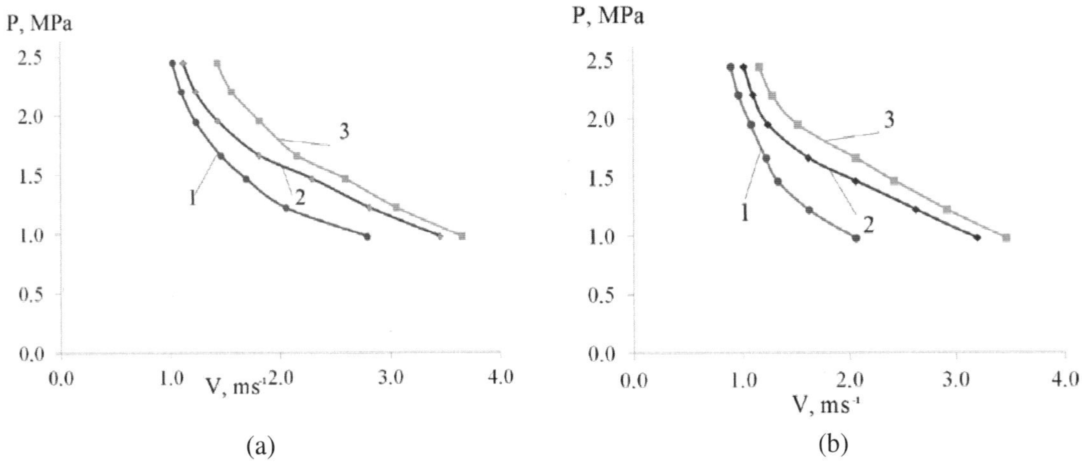

Figure 4. Dependence of the load on the sliding speed and lubricant: a-mineral oil of the "A" brand, 1-bronze 12% Sn, 2-bronze 12% Sn and 10% vol. GE, 3-bronze 12% Sn and 10% vol. GP, b-synthetic oil ATF-III-1-bronze 12% Sn, 2-bronze 12% Sn and 10% vol. GE-1, 3-bronze 12% Sn and 10% vol. GP-1.

4. Discussion

The obtained results of tribological tests show an increase in the jamming load when using mineral and synthetic oil compared to samples without graphite. The size of graphite particles also affects the results of wear intensity 2 times lower on mineral oil when using graphite GP. The friction coefficients are significantly lower when using synthetic oil. The results obtained can be used in the development of new composite bronze-graphite materials with the addition of pencil graphite.

The use of new materials will improve the reliability of friction units in transport engineering.

5. Conclusion

Studies show that the use of graphite in composite powder materials significantly increases the load of jamming in friction pairs. So at a sliding speed of 2 ms^{-1}, it increases by 1.5-2 times compared to tin bronze without graphite, with graphite GP and 1.2 – 1.5 times and graphite GE when tested with mineral and synthetic oil, respectively.

References

[1] Hussein M K, Jameel W W and Sabah F A 2018 *J. Eng.* **24** pp 49-59
[2] Xiao-Ming H, Fei G, Lin-Lin S, Rong F and En Z 2017. *J. Tribol.* **139**
[3] Yuan J, Zeng J, Duan Z, Pei W and Chen H 2018 *IOP Conf. Series: Mat. Sci. and Eng.* **397**
[4] Kuksenova L I, Lapteva V G, Kolmakov A G and Rybakova L M 2001 *Test Methods for friction and wear* (Moscow: Intermet Engineering) chapter 3 pp 57–63
[5] Biryukov V P, Gudushauri E G and Fishkov A A 2019 *Pat.* 2683600 (RF)

Development and studies of new rolling screw-type mechanisms

D S Blinov[1] **and M I Morozov**[2]

[1] Bauman Moscow State Technical University, Russia
[2] OOO NaukaSoft Experimental Workshop company, Moscow, Russia

dmitriyblinov@mail.ru

Abstract. One of the tendencies in mechanical engineering is the transition from sliding friction to rolling friction in part joints moving under load, motor assemblies and machinery. For the present time, planetary roller screws (PRS), which are rolling mechanisms, possess the biggest potential as converters of rotational motion to linear motion. However production of high-precision parts for PRS requires special expensive equipment. The emerging world tendency for reduction of technological costs is to develop nutless roller screws (NRS), which is the most complex PRS part to manufacture. Its production requires an additional machine tool or a universal grinding machine for external and internal grinding. The new rolling NRS has been developed in line with this tendency. This paper focuses on the design and operation of the mechanism and on theoretical studies. A test model and a testing rig have been developed to confirm the results of the theoretical studies and to determine operation parameters of the new NRS. The experiments showed that the newly developed NRS loses to PRS in some parameters, while outperforming it in other. It is also simpler and cheaper to produce and after additional studies and tests rational applications for the newly developed NRS can be found.

1. Introduction

The role of machine engineering has been considerably growing recently and one of its main problems is related to friction and wear. By different estimates, about 70% to 85% of machine engineering products break down due to wear. The losses due to friction in developed countries account for 5 ... 6% of the national income, 20 ... 25% of energy produced globally in a year is used to overcome friction. With understanding of the importance of this problem for the humanity, the science of friction, wear and greasing, tribology [1], has been intensively developing starting from the middle of the last century, together with its application for engineers, triboengineering [2].

One of the radical means of reducing the negative consequences of friction and wear is development of assemblies and mechanisms employing rolling friction.

The objects of research are roller screws, which have not been widely used in machine engineering in Russia yet, but the need for them is constantly growing. Screw-based rolling mechanisms can be of two basic designs: ball screws and roller screws [3]. One of the roller screw subclasses are planetary roller screws (PRS), figure 1, which, for the present day, are the most prospective screw mechanisms. The screw, rollers and the nut have a special thread with the angle of $\alpha=90°$, figure 2.

PRS in figure 1 consists of multi-start screw 1 and nut 2, n threaded rollers 3, separators 4, and casing parts and sealings, which are not shown in figure 1. In order to synchronize the operation, each roller is additionally connected with the nut by cogged engagement. The profile of the roller thread turns is convex enabling an initial point contact of the roller thread turns with the interfacing thread turns of the screw and the nut and excluding unwanted edge contacts.

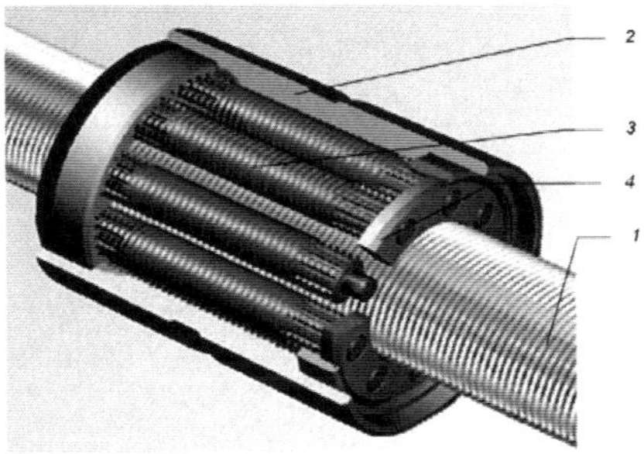

Figure 1. Planetary roller screw.

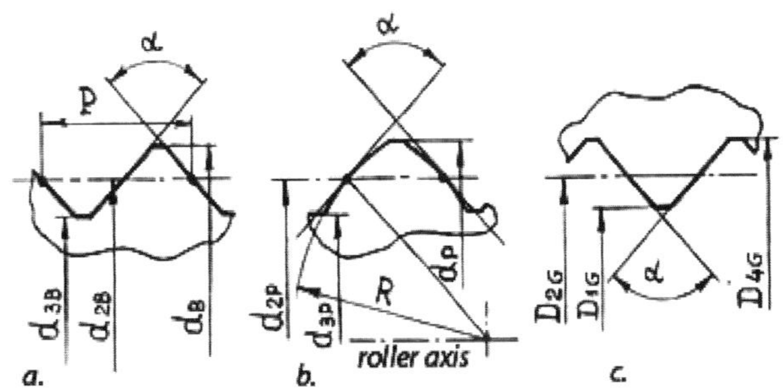

Figure 2. Thread profile: *a* – screw; *b* – roll; *c* – nut.

In order to implement rolling friction in PRS, the pitch diameters of the screw, rollers and nut threads are related to each other and to other thread parameters by mathematically-defined relations [4].

PRS are high precision mechanisms with *n* power flows, closed dimension chains and numerous redundant constraints. Launching of PRS serial manufacturing requires huge material and time input with an unclear prospect. Only expensive specialized machine tools, well-adjusted technology process and machine-tool operator's experience can ensure precision in manufacturing of the main PRS threaded parts. Manufacturing of high-precision multiple threads on the nut, tempered to a high hardness, represents a particular challenge.

The desire to do without the most complex in manufacturing PRS part, the nut, has led to development of nutless roller screws (NRS). The number of NRS patents sold in the advanced industrial countries exceeds several hundred [5-8], that means a tendency for creation of such mechanisms has emerged and we can talk about a new subclass of roller screws – NRS. This confirms the relevancy of their development. It should be noted that NRS should retain one of the PRS most important features that is the rolling friction.

Many of the developed NRS have considerable shortcomings, due to which they are not applicable in practice. In many of the mechanisms rolling friction is not carried out. But new NRS are being developed so that the subclass of these mechanisms is constantly growing.

2. Purpose and objectives of the research

The purpose and objectives of the research are to develop a new operable NRS design, to prove that rolling friction is effected in it, as well as to demonstrate by calculations and tests that the main parameters of the new NRS are similar to the parameters of traditional PRS. Furthermore, NRS should be cheaper and easier than the latter to manufacture, and launching their production, both serial and piecework, should be affordable for many of the Russian enterprises.

3. Studies of PRS friction zones

In order to compare NRS with the prototype (PRS) let's review the main sources of friction losses in PRS under load. These include numerous (hundreds) contact spots of interfacing nut and roller threads, figure 3, and interfacing roller and screw threads, figure 4.

The roller, figure 3, is rotating in planetary movement around the screw axis (point O) with angular velocity of translation motion ω_e, while all the points of the roller are rotating around its axis with angular velocity of translation motion ω_r. When summing up the linear velocities of these two motions only in the initial contact point A, the absolute velocity will be zero, since the roller is engaged by a gear ring with the nut, which does not rotate. When under load, this will cause spin friction around A point in the contact pattern [9].

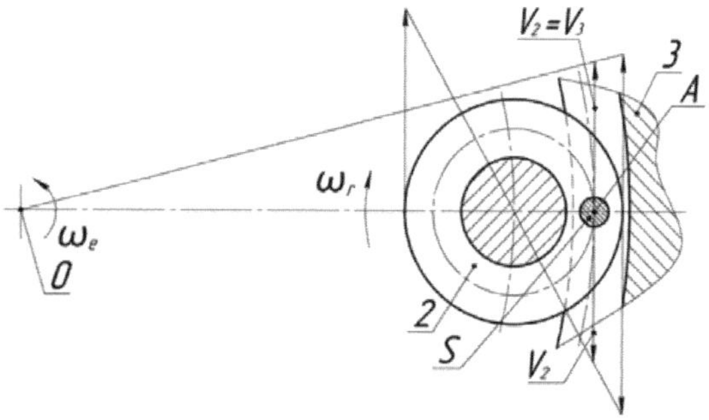

Figure 3. Contact pattern S between thread turns of a roller 2 and a nut 3.

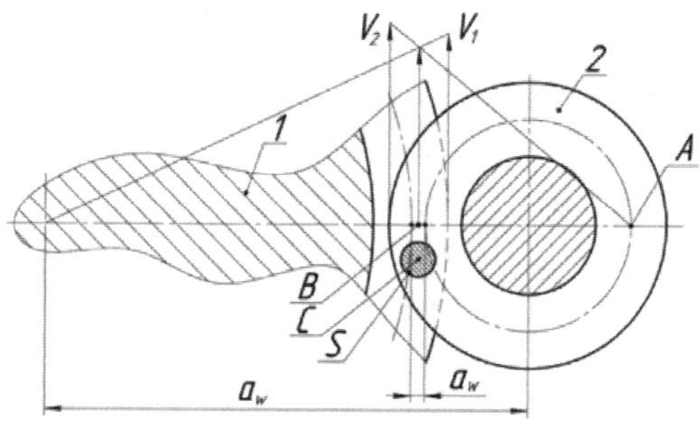

Figure 4. Contact pattern S between thread turns of screw 1 and roller 2.

One of PRS features is the unequal screw and nut thread angles. Due to this, if the PRS center distance is nominal, roller threads don't fit into the cavities between the adjacent screw threads [4, 10]. In order to fit the roller thread into the cavity between the adjacent screw thread turns, it is necessary to increase the nominal value of the center distance by Δa_w, figure 4 [10]. This causes the initial contact point to move to point C relatively to the line that connects the axes of the screw and the roller, figure 4. This line includes point B, in which the velocities of the screw and the roller are equal.

When under load, a contact pattern (area) S will emerge there with point C inside, figure 4. Depending on the load and screw and roller thread parameters, point C may fall either within or outside of area S. In any case there shall be spin friction around point B [9].

4. Development of a new NRS design

PRS nut has two the main functions, figure 1: the nut takes up the load from the screw and transfers it to the casing; the nut keeps the rollers from radial movement. Taking in account these functions and NRS development experience, several NRS have been developed and patented in Russia successively. As a result, two most successful designs were developed [11, 12], one of which (see patent [11]) is shown in figure 5.

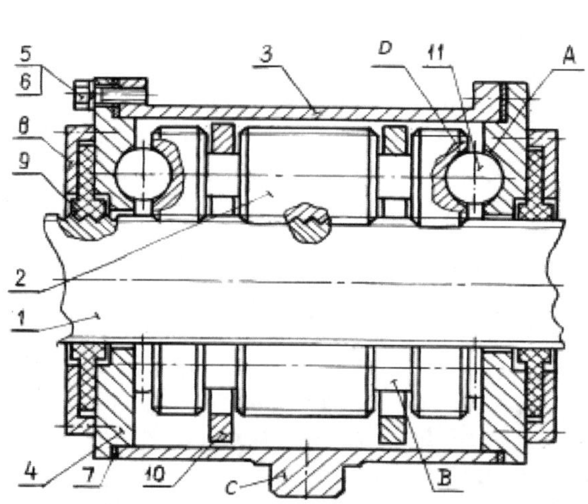

Figure 5. The developed NRS structure with fixed roller axes.

The NRS consists of a screw 1 with $z_в$ thread starts and thread pitch P, threaded rollers 2 with z_p thread starts, installed inside casing 3 with covers 4. The covers are fixed on the casing by the means of screws 5 and lock washers 6. The screw and rollers are engaged by their threads along the screw generators. Adjusting shims 7 are placed between casing 3 and covers 4. Ball seats A with radius $R_к$ are made on the inner face surface of each cover. The number of seats on each cover is equal to the number of rollers. Casing 3 has basic elements C, designed to fix NRS with actuating device. Oil slinger 9 is fixed on the outer face surface of each cover 4 by the means of bushing 8. Each roller not far from its ends has two ring grooves B with installed rings 10. End faces of each roller have spherical seats D with radius $R_p = R_к$. Balls 11 with radius $R_S<R_к$ are installed in the spherical seats A of covers and D of rollers at the same time. To make the axis of each roller parallel with the screw axis, spherical seats A on the covers are made with high precision of angular position, and an angular orientation of covers relative to the casing is ensured by the means of pins.

When NRS is in operation, the screw rotates around its axis, the rollers rotate around their axes and together with the casing make forward motion along the crew axis. The working axial force is transferred from the screw to the rollers via the interfacing thread turns of these parts, then from the rollers to the casing via the contact of spherical seats D on the rollers with the balls and the balls with spherical seats A on one of the covers. Rings 10, which keep rollers from radial movement, rotate.

One important difference of NRS from PRS is the lack of mathematical dependency, which make the basic parameters of the mechanism interrelated, that allows for higher flexibility in designing of NRS. For example, the screw can be single- or double-threaded.

5. Kinematical calculations of NRS

Let's review the NRS velocity diagram, figure 6, with consideration of slipping. Let's introduce a relative slip factor $\zeta = (V_в - V_p) / V_в$, where $V_в$ and V_p are linear velocities of the initial contact point of the interfacing screw and roller thread turns.

Taking the screw angular velocity $\omega_в$ as initial, let's define the angular velocities of the roller ω_p and the ring $\omega_к$

$$\omega_p = \omega_в \cdot (d_{2в} / d_{2p}) \cdot (1-\zeta) \quad (1)$$

$$\omega_к = \omega_в \cdot d_{2в} \cdot d_S \cdot (1-\zeta) / [d_{2p} \cdot (d_{2в} + d_{2p} + d_S)] \quad (2)$$

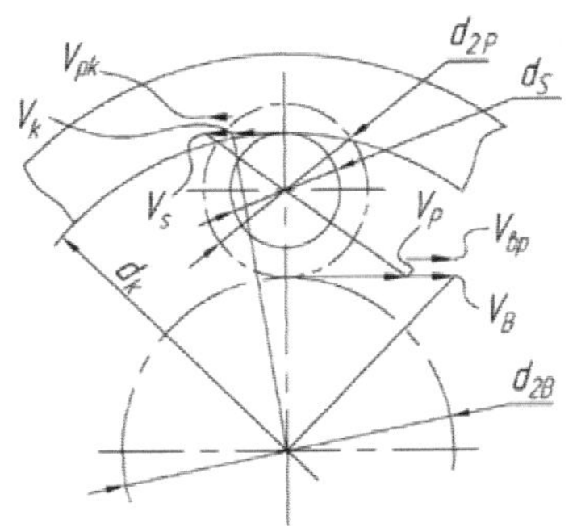

Figure 6. NRS velocity diagram with consideration of slipping. Where: d_S and $d_к$ is the roller pin diameter and the ring hole diameter; V_S and $V_к$ are linear velocities of the roller pin and the ring; $V_{вp}$ and $V_{pк}$ are relative linear velocities.

NRS gear ratio, mm/revolution

$$H = P \cdot z_в \pm (d_{2в} / d_{2p}) \cdot P \cdot z_p \cdot (1-\zeta) \quad (3)$$

In equation (3): «+» is used for unidirectional screw and roller threads; «–» is used for screw and roller threads of different directions.

The effects in the interface of the screw and the roller threads will be the same as in PRS, figure 4. The rings, driven by the roller pins, will be rolling over them with $\omega_к$, while each ball in the spherical seats of the rollers and the cover will orbit around the roller axis, so that spin friction will be carried out in the power contacts. Therefore the developed NRS belongs to screw rolling mechanisms.

6. Theoretical studies of NRS under load

The purpose of the studies is defining the permitted axial force $F_{\Sigma A}$ acting on the mechanism, figure 7, and its operational life. Load capacity and operation life of ball screws and roller screws is accessed by the means of static C_{0A} and dynamic C_A load carrying capacities [13, 14].

Let's assume that all rollers are loaded equally. Figure 7 shows: screw 1, roller 2, two balls 3, ball supports 4 and two rings 5. The following is indicated in figure 7: F_{AP} is axial force acting upon the roller; F_{K1} and F_{K2} are radial forces acting on the 1st (left) and 2nd (right) rings; $F_N[J]$ is normal force, acting on J-th turn of the roller thread. The number of thread turns on roller generator m, where m is of several dozen.

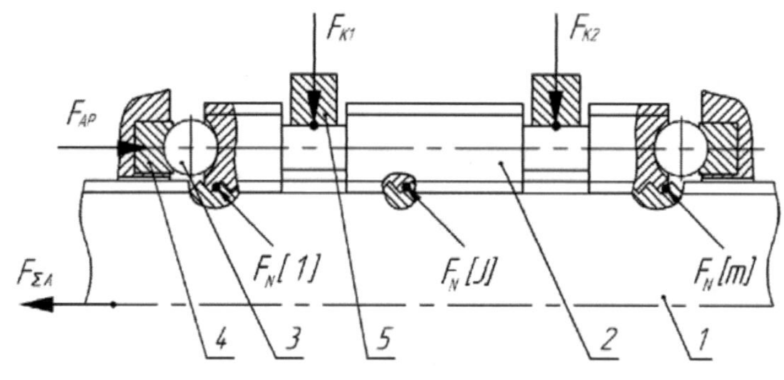

Figure 7. NRS loading diagram.

The specific feature of roller loading is generation of an overturning moment from a pair of forces F_{AP} with the arm (d_{2p} /2). A mathematical NRS model [15] is developed to eliminate redundancy, which considers overturning moment, acting upon the roller, contact flexibility of the interfacing screw and roller thread turns and the ring tension and bending flexibility.

In order to implement the mathematical model a numerical method with forward and backward elimination and a special computer program [15] were developed.

These allows defining the most heavily loaded thread turns of NRS parts, the most heavily loaded junction of a ring with a roller pin, figure 8, and contact force in the ball - spherical seat junction, figure 9.

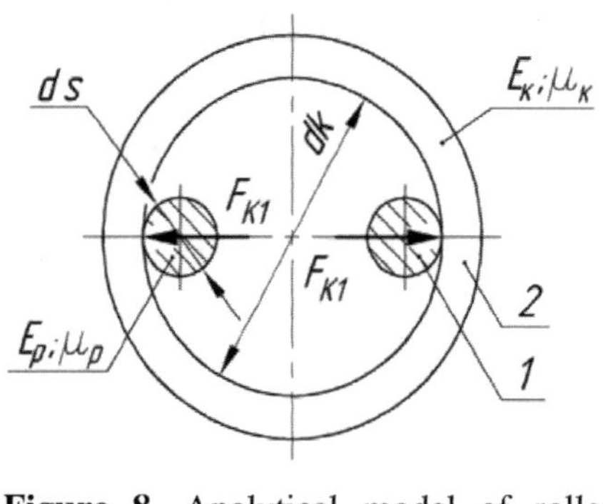

Figure 8. Analytical model of roller pins 1 contact with the ring 2.

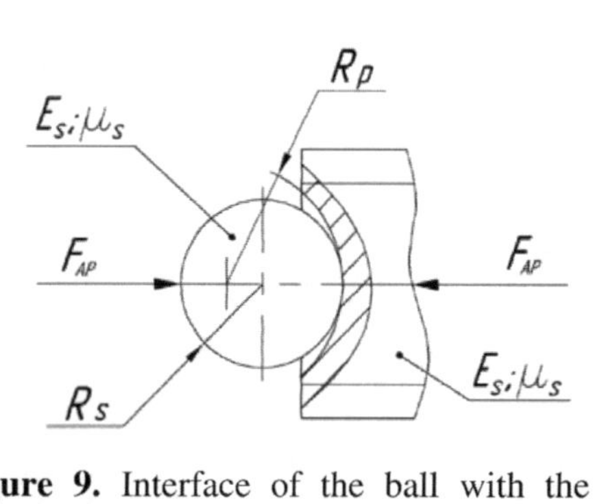

Figure 9. Interface of the ball with the spherical roller seat.

E and μ shown in figure 8 and figure 9 are elastic modulus and Poisson's ratio for the correspondent part.

After defining a pair of the most heavily loaded screw and roller thread turns and the normal force, acting in their junction [15], let's calculate the stress-strain behavior of the thread turns using the Hertz method for an initial point contact [16]. Let's define the stress-strain behavior for the heaviest loaded ring, figure 7, [15] with consideration of tensioning and bending [17] and calculate the stress-strain behavior for the ring, figure 8, using the Hertz method for an initial point contact [16]. Let's define the stress-strain behavior, figure 9, for the initial point contact of the ball and the spherical seat [16].

The biggest stresses generated in these joints or in the most heavily loaded ring define he permitted static strength of the mechanism or its static load capacity C_{0A}. By varying dimensions it is possible to achieve static uniform strength of the parts in the joints.

By using the found correlation between the roller screw load carrying capacities C_{0A} and $C_A$and the number of screw threads z_B, its pitch diameter d_{2B} and thread pitch P, a method for forecasting of dynamic load carrying capacity (life time) of these mechanisms based on the known static load carrying capacity C_{0A} [15]

$$C_A = C_{0A} /[0,3638 \cdot (d_{2B})^{0,5729} \cdot (P \cdot z_B)^{-0,2027}] \quad (4)$$

7. Experimental analysis of NRS

An electromechanical drive, an NRS test model, figure 10, and a test rig, figure 11, were developed and manufactured for the tests. All parts of the test model are manufactured on home-produced equipment [15].

Figure 10. Test NRS model assembly.

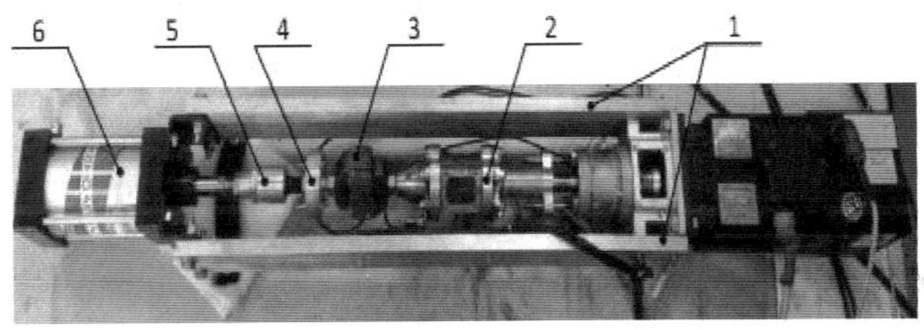

Figure 11. Power part of the test rig.

The test NRS model assembly, figure 10, consists of a screw and a set of rollers, which are fixed to the screw by two rings.

The power part of the rig, figure 11, consists of an electromechanical drive based on NRS 2 and a pneumatic cylinder 6, fixed to the frame 1 and connected to each other via a force sensor 3, a coupling 4 and a flexible coupling 5 (MK–PC080).

The results of theoretical studies on measuring the movement of the ring outer surface points, located approximately at the same radius with the roller axis [15], were experimentally proven.

The assembled and adjusted NRS without greasing and loading was studied in order to define kinematic parameters [15]. The difference between the theoretical and experimental values of the gear ratio is 3%.

A tendency for increasing the operation speed of electromechanical drives, and, consequently, their actuators emerged in the recent years. During NRS tests, the value of the output element forward speed of 379.4 mm·s^{-1} under the maximum possible motor shaft rotation rate of 2036.0 0 rev·min^{-1} (the nominal motor shaft rotation rate is 2000.0 rev·min^{-1}) was achieved [15]. The received rate value is not the limiting value for this mechanism. Defining the maximum possible linear velocity of NRS output element requires continuing the tests with a more high-speed motor.

The maximum value of efficiency factor in the experiments is 81.4 % under the load of 2.0 kN and with the screw rotation rate of 50.0 rev·min^{-1}, the mean value of efficiency factor is about 76%. It should be noted, that these efficiency factor values include losses in the whole drive, consisting of: a roller screw, bearing units, guides, sealings, junctions couplings, anti-rotation device etc. [15].

In order to define the rotating resistance moment of the NRS test model [15] a cold chamber NZ-250/70 and a warm chamber KT-0.05/315M were used. The drive was kept in one of the specified chambers with the specified temperature within 2 hours, after that it was taken from the chamber for testing, figure 12. The tests showed that NRS is operable under the temperatures in the range of from –60°C to +80°C.

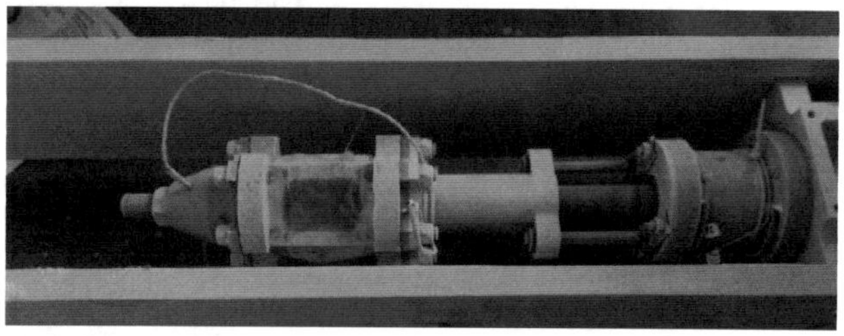

Figure 12. Drive with NRS under the temperature of –60°C.

Life time tests for NRS drive have been conducted in laboratory conditions at a test rig, controlled by LabVIEW and a computer, during one and a half years [15]. During this time the test model was in operation for more than three thousand hours. This corresponds to approximately 1.8 mil. cycles of back and forth movement of the NRS output element. The load to the output element was about 2000 N ± 200 N. The operation (no-load) stroke was 70 mm. The load and the procedure were controlled by LabVIEW and a computer.

Grease «Plasgir» as per STO 07548712-008-2010 was used for the tests. The grease has not been changed during the tests. A range of parameters was measured during the life time tests. For example, the steady-state temperature of the NRS test model depended from the season and temperature of the environment and in different days was from +32°C to + 47°C. A bellows coupling failed and was changed during the tests. By the end of the tests the temperature in the place of double-sided angular-contact ball bearing started to grow and a knocking sound appeared, that means the support devices, as in PRS case, make problems for the drive. Inspection of the NRS test model parts showed, that there were no serious wear of threaded parts. NRS stayed operable.

The produced research proved the operability of NRS test model under a wide range of loads, linear velocities of output element and temperatures.

The paper [15] contains comparison of the developed NRS parameters with the parameters of the traditional roller screws is given in the paper. The values of some of the compared parameters are similar, for some parameters NRS is better than a roller screw, while for the others it is not. There are some parameters, which can't be compared due to lack of information, since there are no data either for a roller screw, or, in the most cases, for NRS. Therefore the further studies of this new mechanism are required.

8. Conclusion

Production of threaded parts for the traditional roller screws requires special expensive high-precision equipment. Threaded parts for NRS can be manufactured on correspondent high-precision equipment, including the universal one. There is no need for a machine tool for grinding a thread on the inner surface of the nut. This is the main advantage of these mechanisms.

For the present time the developed NRS can be used in nonvital products under low and medium loads. Defining the rational fields of application of the developed NRS in vital products of machine engineering requires additional research.

References

[1] *Friction, Wear and Lubrication (Tribology and Tribotechnics)* 2003 ed A V Chichinadze (Moscow: Engineering) p 576
[2] Garkunov D N 1985 *Tribotechnics* (Moscow: Engineering) p 424
[3] Turpaev A I 1982 *Screw Mechanisms and Transmissions* (Moscow: Engineering) p 223
[4] Kozyrev V V 2004 *Design of Roller Screw Gears and Methods of Their Design* (Vladimir: Vladimir state University) p 102
[5] Stanley R 1977 Pat. of the GB No 1490113 *Linear actuator*
[6] Stanley R B 1977 Pat. of the US No 4033194 *Synchronized linear actuator*
[7] Gatrner R 1982 Pat. of the US No 4322987 *Screw drive*
[8] Isamu T et al LTD 2006 Pat. of the Japan No 2006329258 *Rotation-linear motion mechanism*
[9] Kilchevsky N A 1977 *Course of Theoretical Mechanics* vol 1 (Moscow: The science) p 456
[10] Ryakhovsky O A, Sorokin F D and Sokolov P A 2011 Calculation of radial displacement of screw and roller axes and position of a contact point of the screw and roller thread in a planetary screw roller train *BMSTU J. of Mechanical Engineering* **6** pp 7-14
[11] Blinov D S, Kondrashova G P, Ryakhovsky O A, Sokolov P A and Laptev I A 2009 Pat. of the Russian Federation No 2374527 appl. 25.04.2008, publ. 27.11.2009
[12] Blinov D S, Laptev I A and Frolov A V 2012 Pat. of the Russian Federation No 2463500
[13] SKF 2014 *Roller screws* (SKF Group) p 136
[14] Exlar 2017 *Engineered Product Catalog* p 178
[15] Morozov M I 2018 *Method for Determining Rational Parameters of Power Roller-Screw Mechanisms Without a Nut* (Moscow) p 155
[16] *Strength Calculations in Mechanical Engineering. In 3 volumes* 1958 ed S D Ponomarev vol 2 (Moscow: Mashgiz) p 974
[17] *Strength, Stability, Vibrations. Handbook in 3 volumes* 1968 ed I A Birger and Y G Panovko vol 1 (Moscow: Engineering) p 832

IOP Conf. Series: Materials Science and Engineering **996** (2020) 012007 doi:10.1088/1757-899X/996/1/012007

Research on slipper bearings of pumps of reactor systems with lead-bismuth coolant at relatively low temperatures

T A Bokova, P A Bokov, A R Marov and N C Volkov

Nizhny Novgorod State Technical University n. a. R. E. Alekseev, Nizhny Novgorod, Russian Federation

marov_2011@mail.ru

Abstract. The academic focus of the article relates to nuclear power engineering, reactor plants (RP) with heavy liquid metal coolant (HLMC) – lead and eutectic lead-bismuth. The article's purpose is to set out the goals and objectives of the research of slipper bearings for HLMC transfer pumps and to identify the problems of design and operation under these conditions based on the available experience. This experience applies both to energy reactors with high HLMC temperatures (400-550°C), and partly to acceleration-controlled systems and research reactors with HLMC temperature of about 200-350°C. The choice of the type of slippper bearing is justified for further researches of its operability in conditions of low HLMC temperatures.

1. Introduction

Currently, various RP applications are being developed in the world, including the development of acceleration-controlled systems and research reactors with a HLMC temperature of about 200-350 °C. There are also promising projects of power reactors with a coolant temperature at various points of circulation circuits from 300°C and above.

Pumping equipment plays one of the most important roles in RP circuits. Our country has unique experience of creation and operation of the main circulators (MC) of the nuclear submarines (NS) with the reactors cooled by the lead-bismuth coolant of projects 645, 705 and 705K [1]. MCP of reactor plants (RP) of projects 645 and 705K had lower hydrostatic bearings, the operation of which was ensured, inter alia, by the operation of auxiliary pumps (leak return pumps) of the reactor circuit. The MCP of nuclear submarines of project 705 was made with a cantilever arrangement of the shaft and the lower slipper bearing did not have. There is also extensive experience with hydrostatic bearings in the MCP of reactor plants BOR-60, BN-600, BN-800, where they have proven themselves well.

In the open publications, information on hydrostatic bearings operating in liquid metal coolant is insignificant and refers either to alkali metals (sodium, etc.) or to lead coolant.

In Soviet literature, the most detailed hydrostatic bearings, including for sodium pumps, are described in the work of V.M. Budov [2].

An important condition for normal operation of sliding bearings in HLMC conditions is the observance of optimal purity of coolant from impurities. Currently, there is a technology of lead-bismuth and lead coolant of RP, developed, among other things, with the participation of Nizhny Novgorod State Technical Engineering University [1]. This HLMC technology provides operability of power contours of critical reactors of division of heavy-nuclei at operating temperatures of heat carriers 400-550°C. The main provisions of this technology are as follows:

1) Cleaning of coolant oxides (one of their main operating impurities) by treatment of coolant and circuit with reducing gas-steam mixtures.
2) Ensuring the resistance of the structural materials of the circuit due to the formation and pre-formation of oxide protective coatings on their surfaces.
3) Control (one way or another) of oxygen content in the coolant to ensure formation and deformation of protective oxide coatings.

However, such coolant transfer technology is not suitable for the conditions of acceleration-controlled systems and research reactors operating at temperatures of 200-350 °C due to the following factors:

1) Cleaning the channels of hydrostatic bearings as well as their operation in design modes is difficult, since the effective reduction of HLMC oxides by reducing gas mixtures will not occur at such temperatures (T = 200-350°C).
2) Formation of oxide protective and antifriction coatings on steel surfaces at T = 200-350°C is very unlikely.
3) It is not possible to effectively control the oxygen content of HLMC by technical means proven at high coolant temperatures for the purpose of cleaning channels and forming the necessary coatings on their surfaces.

Therefore, the team of Nizhny Novgorod State Technical Engineering University aims at experimental development of the most promising type of MCP slipper bearing transferring HLMC under the specified conditions of uncalculated operation. In order to do this, it is necessary to analyze the entire experience of operating bearings in HLMC along with analysis of the experience of the team - to choose the most promising type, conduct experimental studies in various operating modes and develop recommendations on the most optimal design versions of the bearing and its operating modes.

2. Research materials and methods

The specific properties of the high temperature liquid metal coolant affecting the operability of the slipper bearings in the conditions of the reactor circuits are as follows [1, 3, 4]:

- Traditional lubricants cannot be used in reactor circuits with HLMC due to high temperatures and the inadmissibility of contact of coolant with organic compounds. The only medium in which such a bearing can operate is the coolant itself.
- All liquid metal coolants (lead, lead-bismuth eutectic, sodium) of reactor plants have a small viscosity value and cannot be considered as lubricants in the traditional sense.
- HLMC does not wet the working surfaces of slipper bearings - adhesion operation is small; the coolant is not able to be held on contact surfaces. Bearing working surfaces as well as other loop surfaces shall have protective oxide coatings.
- Heat is released in the local contact areas of the friction surfaces, which can be effectively discharged by the liquid metal coolant, which has a favorable effect on the operation of the contact pairs.

In wall areas of structural material contact with coolant, formation of oxide protective coatings and disperse systems including impurities impregnated with coolant is possible (and mandatory). As have the experiments of Nizhny Novgorod State Technical Engineering University showed, they are both antifriction coatings, which has a favorable effect on the operation of contact pairs. The destruction of these dispersed systems and oxide coatings dramatically intensifies the wear process in the contact zone of the working surfaces.

The slip process in a hydrodynamic bearing implies two necessary conditions [5, 6]:

- complete wetting of shaft and bushing surfaces with lubricating fluid;
- high viscosity of lubricating fluid.

In the circuits of lead and lead-bismuth coolant, both these conditions are absent. A necessary condition for the operation of high-temperature circuits with HLMC is the formation and pre-formation of protective oxide coatings (films) on the surfaces of structural materials. The presence of such films ensures the wettability of the oxidized surface of the HLMC, the value of the edge wetting angle of these surfaces in the protective gas medium is [theta] $\approx 110 - 120^{\circ}$. The only medium that can be used as a liquid lubricant in hydrodynamic bearings under the conditions under consideration is a liquid metal coolant. The values of lead and lead-bismuth eutectic viscosity coefficients in the operating conditions of HLMC circuits are small and close to or less than those for water under normal conditions.

It is impossible to create operable hydrodynamic bearings operating in the medium of liquid-metal coolant of reactor circuits (sodium, lead, lead-bismuth eutectics), as confirmed both by design-theoretical and experimental methods [1].

The fundamental incapability of creating hydrodynamic bearings in HLMC, switching to the contact friction mode of hydrostatic bearings during the start-up, shutdown and emergency clogging of bearing channels with impurities particles - initiate interest in studying the operability of contact friction bearings in conditions of circuits with circulation of high-temperature liquid-metal coolants.

By contact (boundary) friction mode is meant a mode in which there is no conventional liquid or plastic lubricant wetting the friction surfaces between the friction surfaces. Under HLMC conditions, it is possible to have a layer of a disperse system (lubricant in solid phase): oxide films on the surfaces of metal structures, as well as particles of oxides or other metal compounds with gas (steam-gas) inclusions dispersed in the coolant of a wall layer impregnated with it. At the same time, periodic or constant washing of contact surfaces with a liquid metal coolant with low viscosity is possible. Under these conditions, direct contact of the "clean" surfaces of the metal surfaces of the structural materials of the contact pairs of the bearing assembly is excluded and heat generated in the contact zone is intensively removed.

Peculiarities of operating conditions of contact friction bearings in HLMC environment include the following:

- mandatory presence of oxide films (coatings) on the surfaces of structural materials of the circuit, which ensure their resistance in the HLMC medium;
- presence of a wall layer with properties different from those of heavy liquid-metal coolant;
- intensive heat removal by liquid metals of heat generated in the friction zone at high operating temperatures of HLMC in the circuit.

The use of contact friction bearings in lead and lead-bismuth coolant reactor circuits seems unlikely in view of their low durability. Their use in bench contours conditions is more attractive in view of the simplicity of the design. The Nizhny Novgorod State Technical Engineering University has experience in their operation as part of bench circuit pumps, which perform the functions of studies other than the studies of bearings (figure 1). Operation of RCP hydrostatic bearings of reactor circuits in contact friction modes is mandatory, a natural necessity, during both the start-up, shutdown and emergency clogging of channels of these bearings with solid-dispersed particles of impurities.

(a) (b)

Figure 1. External surface of shaft bushing (a) and internal surface of contact friction bearing bushing (b) after 950 hours of operation in lead medium at temperature of 480...510°C.

The most promising type of slipper bearing for RP with HLMC operating temperatures of 200-350°C is the hydrostatic bearing (HB) (figures 2, 3).

(a) (b)

(c)

Figure 2. HB of constant throttling (a, b) and shaft bushing (c) of bench pump FT-1 with lead-bismuth coolant.

The Nizhny Novgorod State Technical University has experience in operation of HB of permanent throttling (figure 2) and slot type with double mutual throttling (figure 3) [4]. The first HB is operated as part of the centrifugal pump of the FT-1 bench with lead-bismuth coolant at temperatures of 250-450°C for more than 40 years and the pump operation time for a specific purpose of about one to two months per year. The second HB is operated as part of the axial pump of the FT-4 stand with lead coolant at temperatures of 420-450°C with a total operating time of up to 2000 hours.

Greater wear of the second HB and the shaft bushing is associated with the operation of the axial pump in large ranges of shaft speeds and with a large number of start-stop cycles. The FT-1 stand pump worked most of the time with the standard shaft speed for it. It is also possible that the axial pump of the bench FT-1 did not develop a sufficient static head for its bearing, unlike the centrifugal pump of the bench FT-1. This aspect requires further investigation.

For further investigation, the HB with double mutual slot throttling was chosen as the most promising and having advantages over the HB of constant throttling [2].

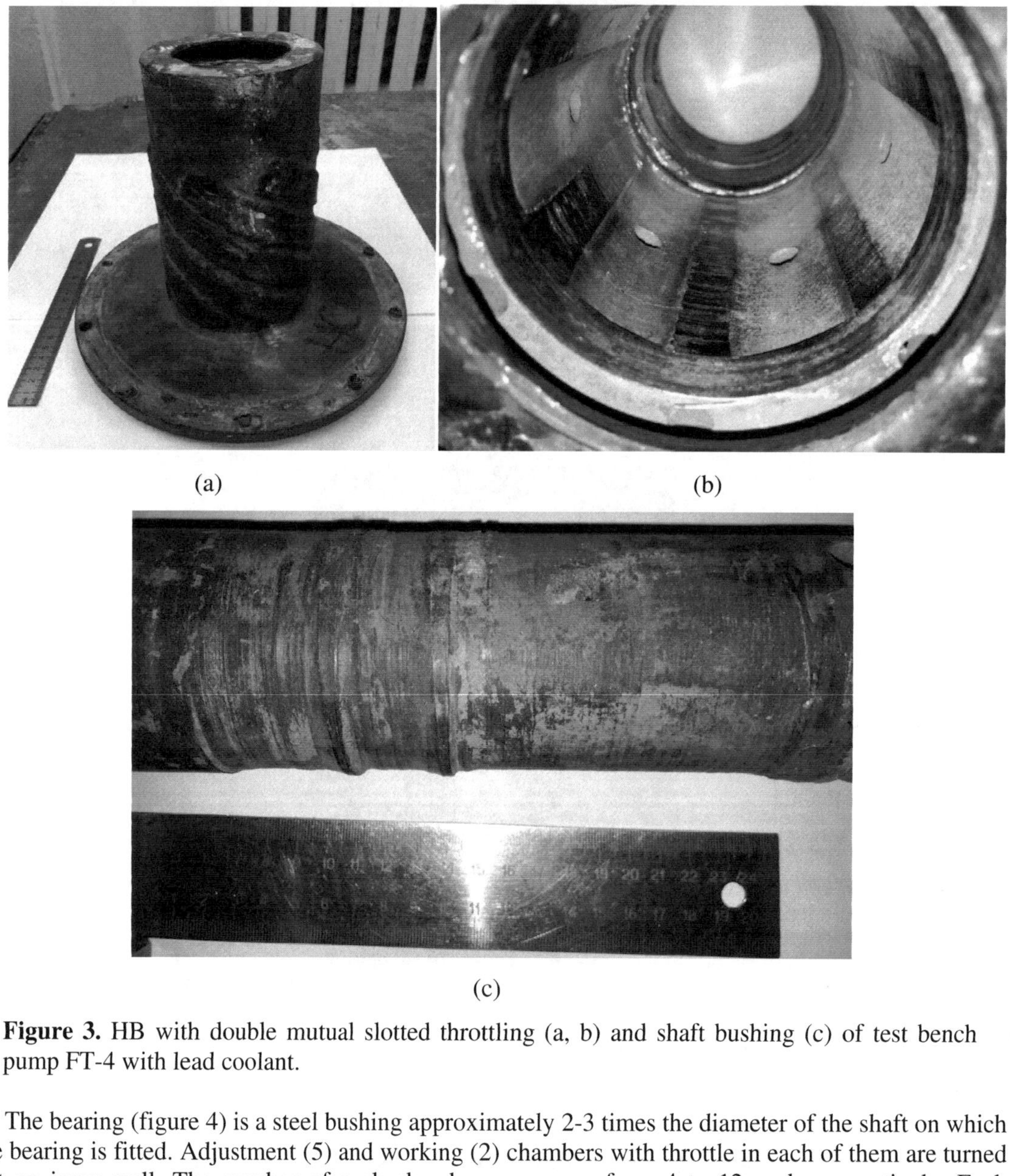

(a) (b)

(c)

Figure 3. HB with double mutual slotted throttling (a, b) and shaft bushing (c) of test bench pump FT-4 with lead coolant.

The bearing (figure 4) is a steel bushing approximately 2-3 times the diameter of the shaft on which the bearing is fitted. Adjustment (5) and working (2) chambers with throttle in each of them are turned out on inner wall. The number of such chambers can vary from 4 to 12 each, respectively. Each adjusting chamber is connected through throttle with groove to each working chamber. These grooves are recessed on the outer wall of the bearing. They can be welded with steel strips or closed by an outer sleeve over the entire bearing. On the inner wall of the bearing, parasitic leakage manifold (1,3) and high pressure manifold (4) are also turned from above, from below and between chambers. Hydrostatic bearing is made of Steel 40Cr13 All-union State Standard 5632-72.

HB tests as part of the experimental site on a research experimental bench with lead-bismuth coolant FT-1 NGTU under various operating modes in the medium of lead-bismuth coolant at a temperature of 200-350°C will allow the bearing to be worked out in conditions of emergency clogging of channels with oxides.

Also, the HB will be tested at various values of misalignment with the shaft and radial loads with real-time recording of heat emissions and braking moment on the shaft. It will also record wear after each test cycle.

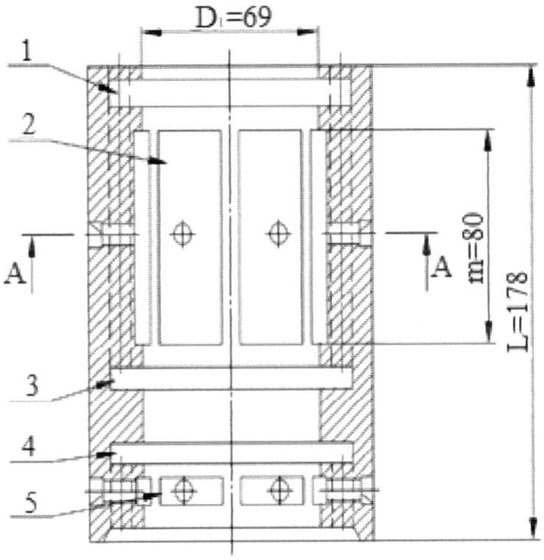

Figure 4. HB model with double mutual slot throttling.
1, 3 – parasitic leakage collector;
2 – operating cameras (6 pcs.);
4 – hp header;
5 – leveling cameras (6 pcs)

3. Anticipated results

Upon completion of the works, it is planned to obtain the following results: the braking moment of the force developed on the shaft, the distribution of the temperature field in the bearing wall, the analysis of splines and the measurement of the change in the roughness of friction surfaces under various operating modes, including after life tests. These results will provide conclusions and recommendations for the design, production and operation of the hydrostatic bearing.

During the tests, it is likely that it is impossible to control the oxygen content in the HLMC at temperatures of 200-350°C, and therefore the operation of the HB with clogged channels is a «dry» friction mode. To overcome this circumstance, it is possible during the experiment to develop a method for maintaining the coolant technology, which is a useful side result.

4. Conclusion

Theoretical and experimental justification of selection of optimal slipper bearing for further experimental investigation in RP conditions with lead-bismuth coolant with temperatures of 200-350°C is presented. At these temperatures, it is difficult to control the oxygen content of the HLMC, which means that the operating mode of the bearing with clogged channels is most likely, which can lead to significant wear and tear. This circumstance makes it necessary to impose more stringent requirements for the reliability and durability of bearing supports and for maintaining the coolant technology.

References

[1] Beznosov A V, Bokova T A and Bokov P A 2016 *Technologies and main equipment of the loops of reactor plants, industrial and research stands with lead and lead-bismuth coolants* (Nizhny Novgorod: Litera)

[2] Budov V M 1986 *Pumps of Nuclear Power Plants*(Moscow: Energoatomizdat Publ) pp 286–329

[3] Beznosov A V, Dragunov Ju G and Rachkov V I 2006 *Heavy liquid metal coolants in nuclear*

energy (Nizhny Novgorod: NNSTU Publ)

[4] Antonenkov M A 2013 *Justification of design solutions for hydrostatic bearings of the main circulation pumps of fast neutron reactors cooled with lead and lead-bismuth coolants* dissertation on PhD (Place of defense Podolsk: JSC OKB Gidropress)

[5] Sinev N M and Udovichenko P M 1972 *Sealless water pumps* (Moscow: Atomizdat)

[6] Voskresensky V A and Dyakov V P 1984 *Calculation and design of sliding bearings (liquid lubrication)* (Moscow: Energoatomizdat)

[7] Beznosov A V, Marov A R, Bokova T A and Volkov N S 2019 Justification of the possibility of creation of sliding bearings, work able in reactor circuits with lead-bismuto and lead heating agents at temperatures of 200-350°C (new technology of HLMC) *XXI Int. Conf. of Young Specialists in Nuclear Power Plants Collection of reports* (Podolsk: OKB Gidropress) pp 164-75

Application of titanium nitride and titanium alumonitride as intermediate layer for diamond-like coatings of steel parts

I A Buyanovskii[1], **V D Samusenko**[1], **V A Levchenko**[2]

[1] Department of Friction and Wear, Mechanical Engineering Research Institute of the Russian Academy of Sciences, 4 M. Kharitonyevskiy Pereulok, Moscow, Russia, 101990
[2] Faculty of Chemistry, Lomonosov Moscow State University, GSP-1, Leninskie Gory, Moscow, Russia, 119991

samusenkovd@gmail.com

Abstract. The physicomechanical and tribological characteristics of DLC coatings was investigated, with using TiN and AlTiN hard coatings as intermediate layers. The tests were carried out both with dry friction and in the environment of PAO-4. It was found that intermediate coatings provide good adhesion of the DLC to steel; the best result was obtained with the AlTiN coating. The main factor determining antifriction properties in the friction pair "steel – coating" with boundary lubrication is tribological properties of coatings.

1. Introduction

Diamond-like coatings (DLC) have good protective and tribological properties. They also have good prospects for application in friction units requiring high reliability and antifriction properties [1], but often due to high internal stresses in diamond-like coatings there is a problem with adhesion to the substrate [2]. For this reason, multilayer coatings are used, which retain their high tribological properties due to the upper layer of diamond-like coating and have good adhesion due to intermediate layers.

Generalized results of tribological tests of diamond-like coatings using titanium nitride [3] and titanium aluminum nitride [4] as intermediate layers are considered in this paper.

2. Materials and research methods

2.1. Samples

The properties of a diamond-like coating were investigated; titanium nitride (TiN) and titanium aluminum nitride (AlTiN) were used as intermediate layers. Cylindrical rollers 8x8 mm (D x L) made of 100Cr6 steel were used as samples for coating deposition. The investigated coatings were deposited to cylindrical surfaces of rollers by the vacuum-arc method of coating deposition (PVD-technology) [5].

Samples with deposited TiN and AlTiN coatings with a thickness ~ 1 μm were made; some of these samples were subsequently coated with a layer of DLC with a thickness of 0.3…0.5 μm. As a result, samples were obtained with TiN and AlTiN coatings, and with two-layer TiN-DLC and AlTiN-DLC coatings.

Measurements of nanohardness, elasticity modulus and adhesion of coatings were carried out on a nanohardness tester NanoScan-4D. Hardness was measured in accordance with recommendations of GOST R 8.748-2011. The Berkovich triangular diamond tip was used as indenter. The maximum loading force was 10 mN (the number of indents in the series is at least 30). Time of loading and unloading was 10 s. The results of measurements of the samples surface characteristics are shown in table 1. It can be seen that the nanohardness of samples with DLC is approximately 4 times higher than that of steel and approximately 2 times higher than that of the coatings that were used as intermediate layers.

Roughness of samples was measured on the optical profilometer S Neox along a profile with length of 100 μm along the generatrix. The results of roughness measuring are shown in table 1. Based on the results, the layers of titanium nitride and titanium aluminum nitride used to adhesion improving have almost no effect on the surface roughness. The application of DLC reduces the surface roughness.

Table 1. Characteristics of samples.

	100Cr6 steel	TiN	TiN+DLC	AlTiN	AlTiN+DLC
Arithmetic average roughness Ra (nm)	86	89	31	80	57
Indentation depth (nm)	184 ± 16	120 ± 24	104 ± 14	138 ± 12	114 ± 12
Hardness H (GPa)	9.4 ± 1,5	22 ± 7	49 ± 13	18.0 ± 2.4	41 ± 9
Elastic modulus E* (GPa)	250 ± 31	400 ± 100	530 ± 110	264 ± 40	440 ± 80

The test procedure for evaluating the adhesion consists in making scratches with varying load and registration minimum force at which the coating delaminate from the substrate (ASTM C 1624). Delamination was determined by an optical image. Each sample was subjected to 3 scratches with a load from 10 mN to 2 N. The length of the scratch was 1 mm. The load was increased linearly. Photographs of scratches on the surface of samples with x10 magnification are shown in figure 1.

On a sample with TiN coating (figure 1 (a)) no delamination of the coating was observed. On a sample with TiN-DLC coating (figure 1 (b)) the destruction of the coating is stably observed at the load of 200 mN. Destruction of the coating takes place in the middle of the TiN layer. Chipping in the titanium nitride layer occurs at the depth of 0.3…1 μm from the surface of the coating.

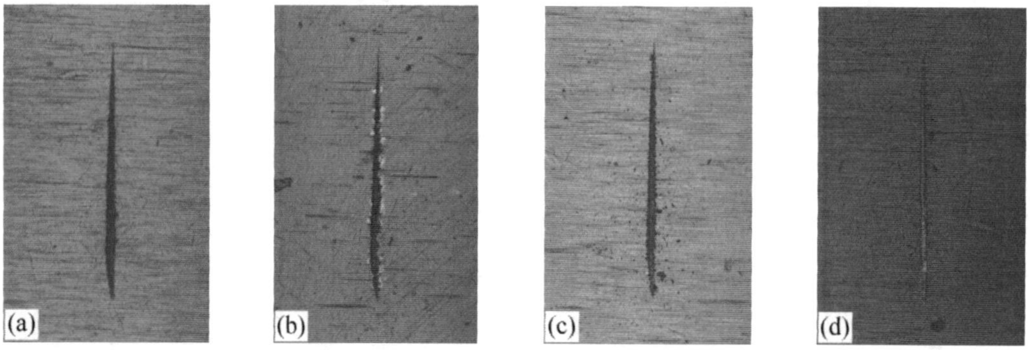

Figure 1. Photographs of scratches on the surface of samples when evaluating the adhesion of coatings: (a) – TiN; (b) – TiN+DLC; (c) – AlTiN; (d) – AlTiN+DLC.

On the sample with AlTiN coating (figure 1 (c)) no delamination of the coating was observed on two scratches. On the third scratch the coating began to delaminate at the load of 100 mN near the edge of the sample, which is probably due to defects during deposition in this zone. On samples with AlTiN+DLC coatings (figure 1 (d)) no delamination was observed. On the surface of the sample coated with AlTiN+DLC visible defects formed during deposition.

2.2. Tribological tests

Tribological tests were carried out on the friction test machine KT-2 according to the "rotating ball - three rollers" friction scheme [4]. Standard balls with diameter of 12.7 mm made of steel "ШХ-15" (100Cr6), and rollers 8x8 mm, on the cylindrical surface of which the investigated coatings were applied as samples for testing. The machine spindle with a ball clamped in it under the load of 108 N is pressed against the cylindrical surfaces of rollers folded into an equilateral triangle, which are installed in a mandrel in a plane perpendicular to the spindle axis. The spindle speed with the ball was 1 min^{-1}. The test duration for each set of samples was 60 min. During the study, values of the frictional moment are recorded, estimated by the angle of the torsion dynamometer twisting. Test results are presented in the form of dependences of the friction coefficient on the test duration. The graphs show the average values of the friction coefficient obtained from the results of three repeated tests. After the end of each test, wear scars were photographed on an optical microscope to assess the degree of damage of the sample surface.

3. Results and discussion

The results of tribological tests of samples without coating and with TiN and AlTiN coatings are shown in figure 2. The microphotographs of wear scars obtained after the end of tests are shown in figure 3.

When tested without lubricant (figure 2 (a)), the antifriction properties of all samples have similar dependences and sufficiently close values of the friction coefficient, taking into account the high scatter of values obtained during dry friction. The friction coefficient begins to increase strongly at the beginning of tests and stabilizes at 0.8…0.9 (steel-to-steel friction coefficient), which is probably a consequence of destruction of coatings (figure 3 (b) and (c)). The TiN coating has undergone destruction to a greater extent and the wear scar exceeds the size of wear of the uncoated specimen, which indicates that the destroyed coating particles act as an abrasive and intensify surface wear.

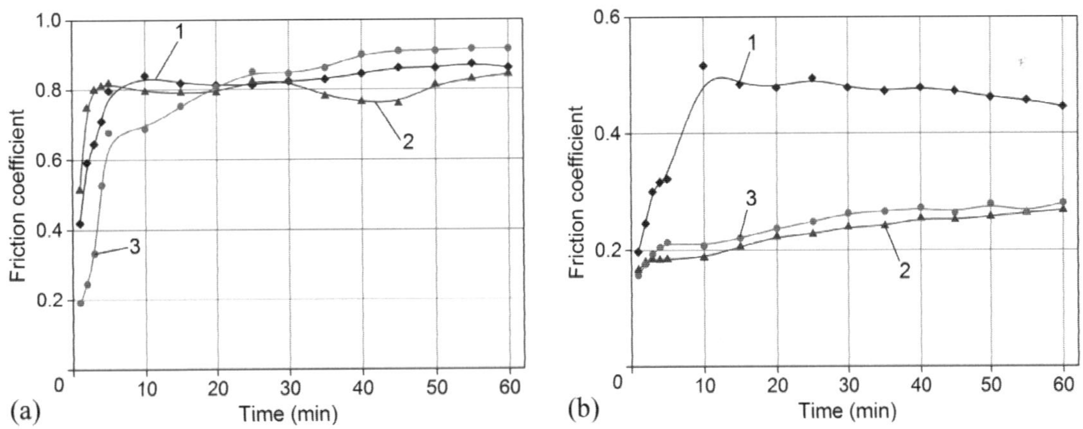

Figure 2. Dependences of the friction coefficient on the test duration without a lubricating medium (a) and in PAO-4 (b) samples: 1 – without coating; 2 – with AlTiN coating; 3 – with TiN coating.

Under test conditions with the PAO-4 lubricant medium (figure 2 (b)), the antifriction properties of samples without and with coatings differ significantly, while the difference in antifriction properties between TiN and AlTiN coatings is insignificant and is within the measurement error. When testing samples with coatings, the coefficient of friction gradually increases throughout the test from 0.16 to 0.28, which is probably associated with gradual deterioration of coatings. As it can be seen from the photographs of wear scars (figure 3 (e) and (f)), the coatings are damaged and traces of plastic deformation of the surface are visible, however, during the test, the destruction of coatings under these conditions does not occur.

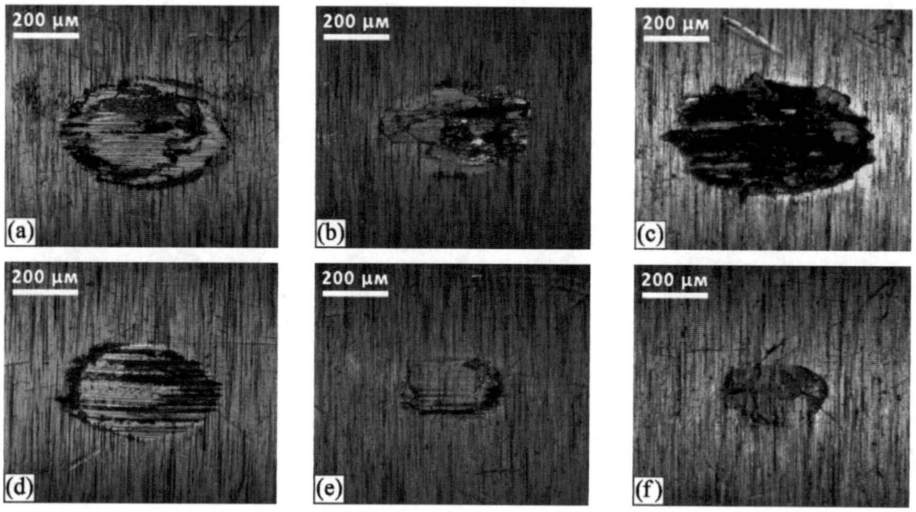

Figure 3. Photographs of wear scars after testing: (a) without coatings and lubricating medium, (b) AlTiN without lubricating medium, (c) TiN without lubricating medium, (d) without coating in PAO-4, (e) AlTiN in PAO-4; (f) TiN in PAO-4.

The results of tests of TiN-DLC and AlTiN-DLC coatings without lubricating medium show (see figure 4), that the friction coefficient with TiN-DLC coating is higher, which is a consequence of the coating destruction. The results of obtained wear scars are presented in figure 5. The increase in the friction coefficient during TiN-DLC test is most likely associated with stages of coating destruction (figure 5 (a)) and friction coefficient tends to the values obtained during testing of uncoated specimens. The AlTiN-DLC coating didn't destruct during the test, only small traces of wear are visible on the surface (figure 5 (b)). Also it has a stable friction coefficient of 0.1 after the running-in stage.

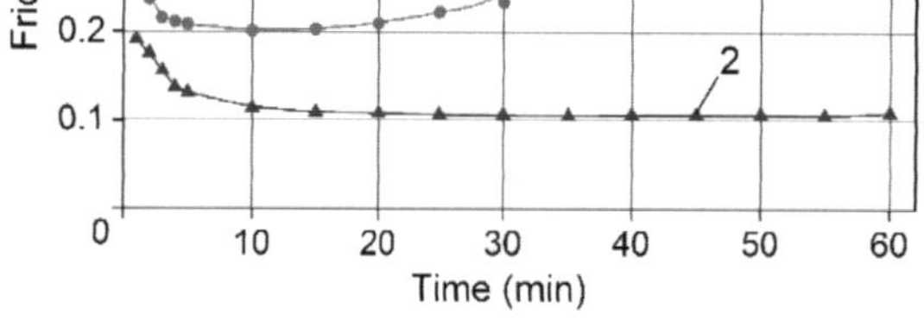

Figure 4. Dependences of the friction coefficient on the test duration of TiN-DLC (1) and AlTiN-DLC (2) coatings without lubricating media.

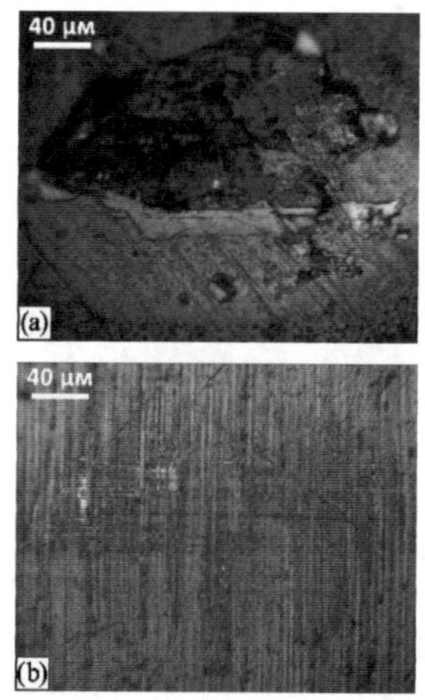

Figure 5. Photographs of wear scars after testing TiN-DLC (a) and AlTiN-DLC (b) without lubricating media.

Results of tribological tests of TiN-DLC and AlTiN-DLC coatings in PAO-4 lubricating media are shown in figure 6. The resulting wear scars are shown in figure 7. Antifriction properties of these two coatings in PAO-4 are close in values at a steady friction mode. At the stage of running-in the friction coefficient decreases and then stabilizes in the range of 0.09…0.11. The running-in process for TiN-DLC coating test is faster, and the resulting wear scar is larger according to the test results (figure 7).

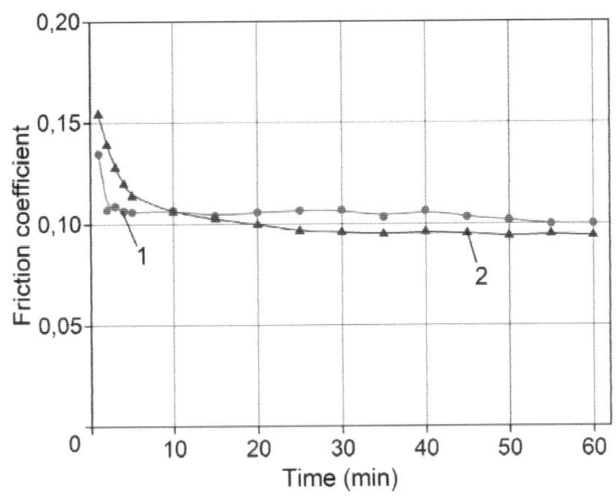

Figure 6. Dependences of the friction coefficient on the test duration of TiN-DLC (1) and AlTiN-DLC (2) in PAO-4.

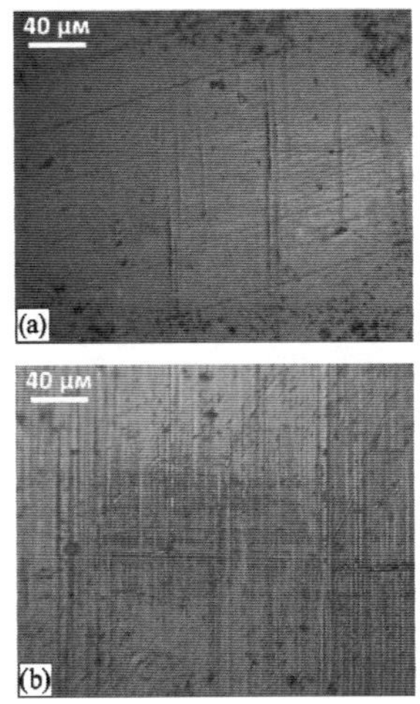

Figure 7. Photographs of wear scars after testing TiN-DLC (a) and AlTiN-DLC (b) in PAO-4.

The antifriction properties of the AlTiN/TiN-DLC coating do not differ significantly depending on the presence of the lubricating medium. From this result, it can be concluded that the presence of DLC is the determining factor for the antifriction properties of the tested friction pair. In [3,4], more detailed results of tribological tests of DLC with intermediate TiN and AlTiN layers in various lubricating media are given.

4. Conclusion

One of the most important characteristics of coatings is adhesion, which allows to maintain the required antifriction properties throughout the required life cycle of the friction unit. The destruction of coatings leads to a growth in the friction coefficient and increases the wear of surfaces in the coupling due to the formed abrasive particles from the destroyed hard coating.

Evaluation of adhesion showed that the AlTiN coating is more preferable than TiN for use as an intermediate layer for diamond-like coatings due to providing better adhesion to the sample surface. A similar result was obtained based on the tribological tests: the AlTiN coating is less susceptible to wear and destruction during friction, which ensures the preservation of the antifriction properties of the diamond-like coating.

The lubricating medium is not the main determining factor of the antifriction properties in the friction pair “steel - DLC”; the most significant in this case is the diamond-like coating itself, and the greatest attention should be paid to the ability of coating to resist destruction.

References

[1] Tyagi A, Walia R S, Murtaza Q, Pandey S M, Tyagi P K and Bajaj B. 2019 *Int. J. Refract. Met. Hard Mater.* **78** 107
[2] Wang M, Schmidt K, Reichelt K and Dimigen H 1992 *J. Mater. Res.* **7** 667
[3] Buyanovskii I A, Samusenko V D and Levchenko V A 2020 *J. Mach. Manuf. Reliab.* **49** 389
[4] Levchenko V A, Buyanovskii I A, Samusenko V D, Matvienko V N and Shcherbakov Y I 2019

J. Frict. Wear **40** 536
[5] Inspektor A and Salvador P A 2014 *Surf. Coat. Technol.* **257** 138

Diagnostics of dissipative characteristics of friction damper

S L Chernyshev[1], V I Kolesnikov[2], V D Vermel [1], A L Ozyabkin[2], I V Kolesnikov[2] and A N Koryakin[1]

[1] Central Aerohydrodynamic Institute, Moscow, Russian Federation
[2] Rostov State Transport University, Rostov-on-Don, Russian Federation

Ozyabkin@mail.ru

Abstract. The aim of our study was to broaden current knowledge of characteristics of friction damper that acts to dissipate the energy of small translational oscillations in friction pairs and to convert it into significant angular shaft movements. We have devised a novel methodology to evaluate elastic-dissipative characteristics of the damper, which provides real-time identification of the stability of these characteristics, as well as running-in periods, stable operation, and catastrophic wear. It is shown that plan bearings made of antifriction self-lubricating organic-fiber composite are more effective than plain bearings made of metal-fluoroplastic tape since their use causes less high-frequency noises that affect the quality of transient response.

1. Introduction

One of the core problems of ensuring a high level of safety in both civil and military aviation is increasing the strength, durability, and damage tolerance of an aircraft structure by diverse methods, including damping technologies for aircraft vibration attenuation. From a technical point of view, the most optimal points for mounting a damper are the points of the damped unit since then its displacements have the highest vibration amplitudes, and the load on the damper is minimized. However, the design features of the damped unit (for instance, podded engines, landing gear assemblies, etc.) do not often meet these requirements. In this case, the only possible points for mounting the damper are the points, characterized by small vibration amplitudes of the structure (for example, near the mounting supports of the damped unit), and large loads on the damper. Under these conditions, the use of well-known devices, in particular hydraulic dampers, appeared to be ineffective. With this in mind, we tried to find an innovative solution.

The research team of Federal State Unitary Enterprise “Central Aerohydrodynamic Institute” (hereafter TsAGI) has developed an innovative type of damper (figure 1) based on sliding friction pairs [1]. Due to the conversion of small translational movements caused by the vibration of the damped structure, the vibration energy of the structure is dissipated into significant angular shaft displacements in the friction pair. It ensures effective damping of vibrations with small amplitudes in a broad range of frequencies. The most important demand for a friction pair is a high level of stability of its tribological properties in the required period of operation.

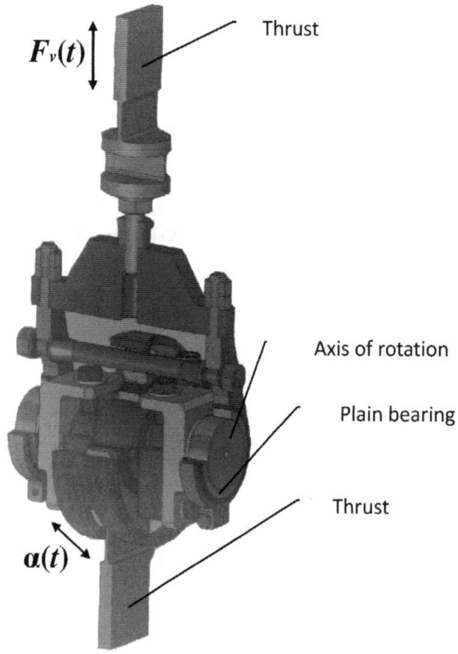

Figure 1. Friction damper scheme.

After the application of load $F_v(t)$, the eccentric shaft moves around the axis of rotation through the angle $\alpha(t)$ and the axis of load application is displaced vertically a small distance from the initial position. With a relatively small eccentricity of the load application, small translational movements of the vertical thrust correspond to large angular displacements of the eccentric shaft with relatively large frictional moments in rotating pairs, which always prevent rotation. As proved, it ensures effective vibration damping in a broad frequency range.

2. Aim of the study

We initiated this research to develop a strong coating for dry friction bearings to be used in a friction damper for reducing vibration of aircraft structures. Besides, we have examined the mechanical properties of coatings and a damper with plain friction bearings to pre-estimate the influence of various technological factors on the tribological properties of coatings.

3. Methodology

We have devised a cutting-edge methodology to diagnose the tribological properties of a damper with rotating friction pairs [2]. It includes the analysis of the known load amplitudes and friction forces, calculation of the frequency transfer function and the transient function of the dynamic coefficient of friction [3]. Moreover, we have employed the analysis of the amplitude-phase characteristics of the dynamic coefficient of friction to determine the dependence of the change in the integral quality criteria on the ratio:

- elastic-inertial components that contribute to bringing up friction pairs together, increasing stresses and temperatures;
- inertial constituents that lead to the loss of stability of the friction bonds;
- inertial-dissipative components of the resistance to relative sliding;
- inertial-dissipative components of resistance, causing frictional self-excited oscillations.

The analysis of the obtained characteristics made it possible to select the recommended values of the dimensionless damping coefficient in the range from 0.08 to 0.38 and its optimal value equal to 0.25. It is worthwhile noting that we have devised a novel methodology of dynamic monitoring of the damper, which allows to keep its technical condition under observation, as well as to predict the degradation of the tribological properties of the materials used.

Experiments on dissipative characteristics of a damper with plain bearings possessing diverse tribological properties were performed on a test bench in 2018 (figure 2). Carrying out experiments, we recorded the amplitude values of the load and the amplitude of the relative displacement at an oscillation frequency ranged from 4 to 15 Hz. In order to analyze the results obtained, we applied the previously developed methodology of dynamic monitoring of the elastic-dissipative characteristics of the damper.

Figure 2. Test bench for conducting experiments on the TsAGI damper.

The dissipative characteristics of the damper were analyzed based on the hysteresis energy loss loop and the method of tribospectral identification of friction processes [4].

4. Results

Figure 3 reports the data on the damper bench tests. It details the analysis of the 2nd, 100th, and 200th loading cycles for three types of plain bearing coatings: a new and worn-out coating made of metal-fluoroplastic tape (figure 3 (a), (b)); a new coating made of antifriction self-lubricating organic-fiber composite (figure 3 (c)). The vibration amplitude of the damper thrust was 0.5 mm at a frequency of $\omega = 4$ Hz. As seen, the damping characteristics of all damper constructions much vary, especially in the area of zero forced displacement amplitudes. A damper with new plain bearings made of antifriction self-lubricating organic-fiber composite (figure 3 (c)) when passing through zero displacement amplitude demonstrates the greatest resistance and, probably, has the highest damping characteristics.

The elastic-dissipative characteristics were assessed based on the analysis of frequency (figure 4), transient (figure 5) and integral functions to support the hypothesis.

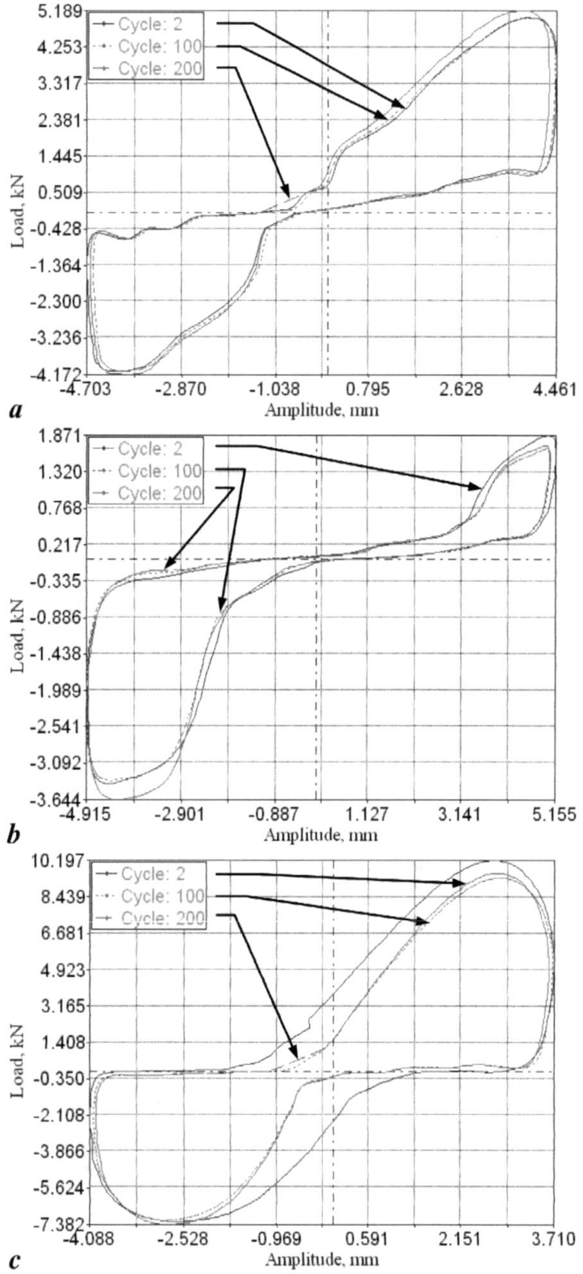

Figure 3. Change in the vertical displacement force of the damper caused by the displacement amplitude of the damper thrust.

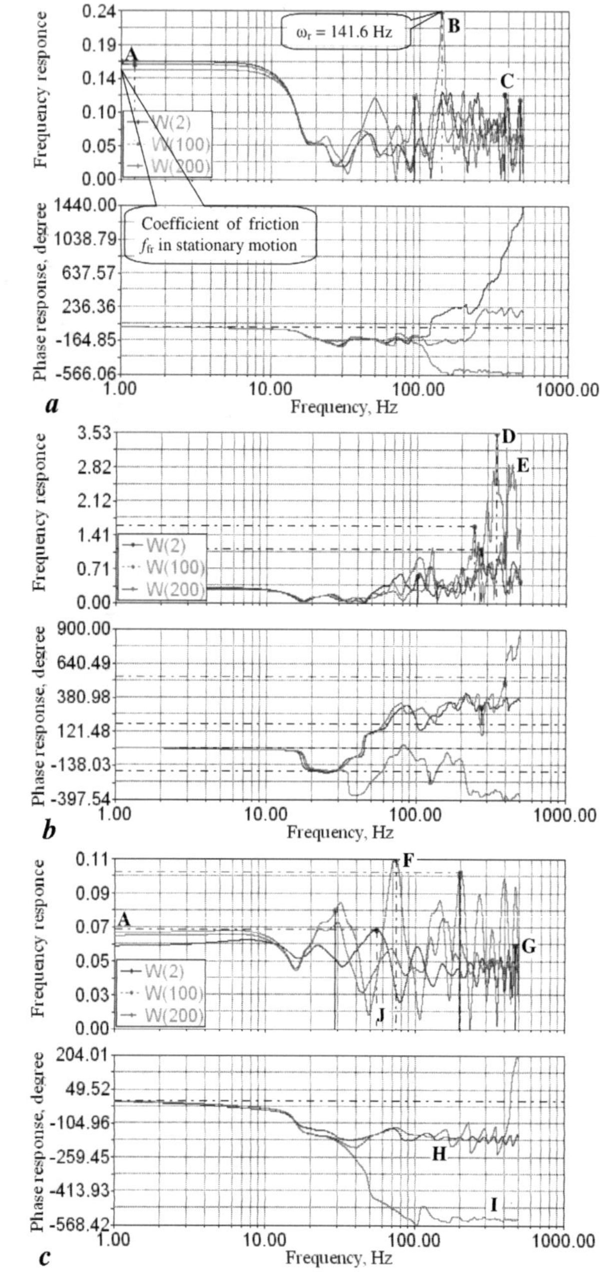

Figure 4. Amplitude and phase frequency characteristics of the damper's dynamic coefficient of friction.

The experiment with new plain bearings made of a metal-fluoroplastic tape (figure 4 (a)) shows that an aperiodic link was effective in suppressing the oscillations in the area of lower frequencies in the frequency range from 0 to 17 Hz. The maximum inertial characteristics are observed at a resonant frequency ω_r = 141.6 and 379.6 Hz with a maximum dynamic coefficient of friction $f_{fr\,max}$ = 0.24 in the 200th and 2nd test cycles (point B and C, respectively). The dynamic coefficient of friction is characterized by the value of gain margin L = 18.3 dB. Significant speeds of relative sliding at higher frequencies lead to the destruction of the metal at the interface of the contact surfaces of rotating friction pairs and notable strain amplitudes of plain bearing bonds.

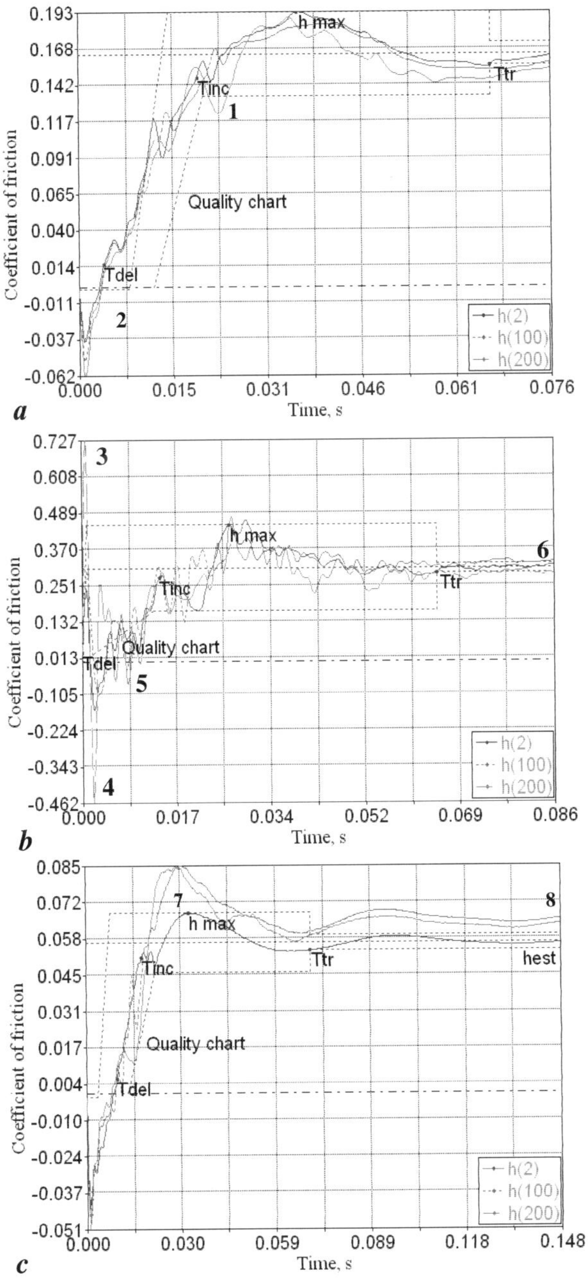

Figure 5. Loading cycling-dependent transient characteristics of a damper.

Figure 4 (b) indicates the unstable operation of a damper with worn-out plain bearings made of metal-fluoroplastic tape, caused by medium (100 ... 130 Hz) and high-frequency (220 ... 500 Hz) vibrations. Along with high values of the dynamic coefficient of friction $f_{fr\ max} = 3.53$ at a frequency of 345 Hz (point D), forces of inertia and resistance (the velocity vector is co-directed with v_{vel}, the velocity vector of friction bond deformation), frictional self-excited oscillations, and significant kinetic energy greatly contribute to the convergence of friction pairs. Moreover, it leads to an increase in contact stresses and the temperature of the friction pairs, the emergence of plastic deformation and

IOP Conf. Series: Materials Science and Engineering **996** (2020) 012009 doi:10.1088/1757-899X/996/1/012009

local areas of thermal-induced seizure. The same dynamic process observed at a frequency of 440 Hz (point E) is, however, characterized with a lower dynamic coefficient of friction f_{fr} = 2.9, the absence of self-excited oscillations, and the formation of athermic seizure bridges.

The plain bearings made of antifriction self-lubricating organic-fiber composite demonstrate the best dynamic characteristics (figure 4 (c)) compared to the other ones. However, the oscillation amplitude of the dynamic coefficient of friction at medium frequencies (in the frequency range from 16 to 120 Hz) increases with the number of cycles (area J). At higher vibration frequencies (from 180 to 500 Hz), in the 2nd and 200th cycles, the dynamic coefficient of friction (area G) does not exceed the coefficient of friction $f_{fr} \approx 0.063$, which corresponds to a stationary stable state. Meanwhile, with an increase in the vibration frequency, the phase argument of the delay happens with seemingly minor deviations (areas H and I).

Obviously, the frequency characteristics affect the quality of the time characteristics in transient responses. Figure 5 presents the transient response to the Heaviside function of a damper.

Large-amplitude high-frequency oscillations of the contacting friction surfaces (point B in figure 4 (a)) seriously affect the deviations 1 of the controlled variable from the stationary motion trajectory (figure 5 (a)) of the corresponding aperiodic link, a significant overshoot level σ, a low negative gradient of the output value, thus increasing the duration of transient response and reducing the effectiveness of regulation.

In the 1st ms of transient response, load and inertia forces in a damper with worn-out plain bearings made of metal-fluoroplastic tape (figure 5 (b)) results in notable convergence of friction surfaces 3 with an equivalent dynamic coefficient of friction f_{fr} = 0.73. In addition, it causes friction seizure that in its turn in the 2nd ms produces stroke-type impulse 4 and deformations in the volumes of contacting friction pairs in the direction of relative sliding v_{vel} (with a coefficient of friction minus 0.46). Under the influence of high-frequency vibrations, until t (h_{max}) = 27 ms the damper is showing maximum deviations from the stationary motion trajectory 5. In the 86th ms, the dynamic coefficient of friction 6 is stabilized at the level of a steady state value f_{fr} = 0.28, which exceeds the permissible value [f_{fr}] = 0.15.

An insignificant dispersion of amplitude harmonics at higher frequencies of a damper with plain bearings made of antifriction self-lubricating organic-fiber composite (see G in figure 4 (c)) considerably improves transient dynamic characteristics (figure 5 (c)). In the 2nd test cycle, after reaching the first maximum h_{max} (*7*), the dynamic coefficient of friction $f_{fr\ max}$ = 0.067 decreases to a steady state value f_0 = 0.058 (*8*). As expected, the quality of regulation of the transient load characteristics F_v is enhanced.

However, it should be noted that frictional vibrations of plain bearings in all dampers lead to the increase in the dynamic coefficient of friction at medium and high frequencies (see areas G ... I in figure 4 (c)), significant overshoot values σ, and duration tp of the transient function regulation.

5. Conclusion

The evidence from the study on the dynamic quality criterion of frequency, time and elastic-dissipative characteristics of dampers with rotating friction pairs points towards the idea that, in contrast to plain bearings made of metal-fluoroplastic tape, the use of those made of antifriction self-lubricating organic- fiber composite allows increasing the stability of damper dynamic and tribological characteristics by 37%.

Under the influence of 1(t)-type stepwise effect and a positive gradient of the rate of frictional links formation at some point in time of transient responses, there is no observation of frictional link deformation, and the dynamic coefficient of friction is equal to 0 (the Tolstoi – Push effect). The subsequent increase in the dynamic coefficient of friction causes the convergence of the contacting friction surfaces, deformation in the active volumes of the friction surfaces, and an increase in contact stresses. Significant speeds of the relative slip in rotating friction pairs cause plastic deformation of the microgeometry of friction surfaces, frictional self-excited oscillations, which lead to the excitation of high-frequency forced oscillations, whose frequency equals the free oscillation frequency, and deviation of the controlled load from the stationary motion trajectory. Due to the athermic or thermic seizure of friction surfaces, the dynamic coefficient of friction tends to infinity (the Tolstoi – Push effect).

6. Recommendations

We have succeeded in finding the alternative solution for suppressing high- and medium-frequency noise during the operation of rotating friction pairs. We have suggested to coat metal surfaces of plain bearings with an antifriction lubricant with excellent adhesive properties before their assembling. The other innovative solution is to metal-clad the surfaces of rotating friction pairs of the damper with high-damping materials. In the case of high contact stresses, these measures will increase the actual contact area of friction pairs and prevent frictional self-excited oscillations, seizure bridges, and fretting corrosion.

Our research was a useful aid when manufacturing of a prototype damper intended for testing as part of a "wing – nacelle" unit of MC-21 aircraft with imitation of the vibrations produced by a running engine.

The methodology for dynamic monitoring of friction systems could be applied to any friction unit of aircraft. Moreover, it has the potential features to make a short-term or long-term forecast of changes in its dynamic characteristics.

References

[1] Baryshnikov O E, Bruskova E V, Vermel V D, Zichenkov M C, Koryakin A N, Puchenkov A L, Paryshev S E, Chernyshev L L and Chernyshev S L 2015 Mechanical Damper of Low-Amplitude Vibrations with Rotating Friction Pairs *Patent 2544046 Russian Federation* pp 14

[2] Kolesnikov V I, Ozyabkin A L, Voropaev A I, Vermel V D and Leonityev A E 2019 Diagnostics of the Dissipative Characteristics of Dampers *Bull. of Mechanical Engineering* vol 3 pp 7-13

[3] Zakovorotny V L and Shapovalov V V 1987 Study on the Complex Coefficient Of Friction *Friction and Wear* pp 22-24

[4] Kolesnikov V I, Ozyabkin A L and Novikov E S 2019 An Innovative Approach to Studying the Processes of Friction, Wear and Monitoring of Heavily Loaded Tribosystems *Friction and Wear* vol 40 No 4 pp 380-88

Acknowledgments

This study was carried out within the framework of State Program of Russian Federation "Development of the Aviation Industry for 2013-2025" and the following research works: "Development of a Methodology for Dynamic Monitoring and Assessment of Elastic-Dissipative Characteristics of a Damper" (Codename "Dragonfly", State Registration Number 22.05.000.001, 2017) and "Improvement of the Current Prototype of the Vibration Damper of Screw Units (SU-5, SU-6) due to Innovative Rotating Friction Pairs" (Codename "Bench-2020", State Registration Number 22.05.000.002, 2018).

IOP Conf. Series: Materials Science and Engineering **996** (2020) 012010 doi:10.1088/1757-899X/996/1/012010

Procedures for measuring thickness of lubricating film using capacitive method on Amsler-type friction machines

G V Darovskoy, V N Polyakov and A N Opatskikh

Rostov State Transport University, Rostov-on-Don, Russia

graffDAROVSKOI@yandex.ru

Abstract. A technique has been developed for measuring the thickness of a lubricating film by the capacitive method on Amsler-type friction machines. A diagram of the implementation of the original device for solving the problem is described, a method for its calibration is described. It was experimentally established that the applied method is highly stable in the results obtained in a wide range of measured values.

1. Introduction

Today, roller friction machines (II 5018, 2070 SMT-1, UMT, etc.) are widely used in tribological tests. At the same time, numerous test experts note the need for their modernization in order to increase the precision of measurements of the parameters of friction modes, as well as to meet the requirements of the latest standards [1-3].

Another vital feature is the choice of lubricant type, as its properties must meet the modern requirements for friction units [4]. Assessing of the operational properties of lubricants requires the knowledge of the lubricant film thickness in the contact zone.

An analysis of the designs of modern Amsler-type friction machines demonstrates that the industry does not make devices for quantitative assessment of lubrication conditions or for measuring film thickness in tribo-couplings. This requires numerous researchers to develop devices of original design.

Snegovsky F.P., Kodnir D.S., Reschikov V.F., Raiko M.V., GuzenkoM.Yu., Pavlik B.B. and others paid much attention to this issue. [5-8].

Their works concern mainly acoustic and electrical (electrical resistance and capacitive) methods of assessing lubrication conditions and measuring film thickness in couplings. This is because these methods have a number of advantages: simplicity of control, convenient data output from an analog-to-digital converter (ADC) to a personal computer (PC) and subsequent processing; low power consumption, small dimensions, etc.

Capacitive methods can be implemented, depending on the shape and size of the parts, in two different methods.

The first is based on the use of a sensor mounted in one of the contacting parts (for example, in tribological monitoring of a "roller-block" friction pair, it is a shaft). The sensor is isolated from the part and serves as a movable capacitor plate; the stationary plate is the opposite area of the second contacting part (the block). In this case, the lubricant is a dielectric.

The disadvantages of this method are nested in the limitations associated with ramming the capacitive sensor into the body of the sample, as well as with the difficulties when using the method on small or complex surfaces (for example, gear wheels, rolling bearings, etc.).

In the second method, contacting parts (the roller and the block) become the capacitor plates, while the lubricating layer acts as a dielectric.

The capacity of such a capacitor can be calculated as the capacity of the flat capacitor

$$C = \frac{S \cdot \varepsilon \cdot \varepsilon_0}{d}$$

where S – area of the block, m^2;

ε_0 – dielectric constant;

ε – relative dielectric permeability of the lubricant;

d – lubricant film thickness, m.

The main disadvantage of this method is the calibration of the sensor.

The authors have experimentally established that many of the proposed methods and original devices for capacitive measuring of the lubricant film thickness have a number of significant disadvantages, such as dispersion in readings, when measuring the lubricant film thickness for rotating roller, even under identical conditions.

2. Goals and objectives of the study

The goal of the paper is to develop an approach for measuring the lubricant film thickness on Amsler-type friction machines.

To achieve this goal, it is necessary to design an original device and a method for its calibration, validate the device, and introduce it into the daily tribological research.

3. Materials and methods of the study

If we include a flat capacitor in a divider (figure 1) and apply to the latter an alternating voltage of high frequency, then the divider's voltage drop will change depending on capacity of the capacitor, i.e. depending on thickness of the lubricant film.

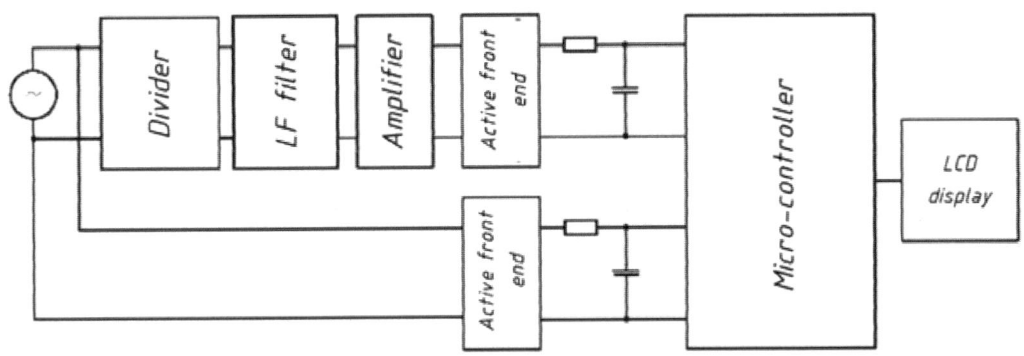

Figure 1. Device for measuring the lubricant film thickness, a functional diagram

The capacitive resistance of such a capacitor can be determined from the dependency

$$x_C = \frac{1}{2 \cdot \pi \cdot f \cdot C} = \frac{d}{2 \cdot \pi \cdot f \cdot S \cdot \varepsilon \cdot \varepsilon_0} \quad (1)$$

where f – current frequency, Hz.

Divider current

$$I = \frac{U_{in}}{2 \cdot R + \dfrac{d}{2 \cdot \pi \cdot f \cdot S \cdot \varepsilon \cdot \varepsilon_0}} \tag{2}$$

where U_{in} - voltage at the divider input (see figure 1), V;

R – divider resistor resistance, Ohm.

Considering (1) and (2), it is possible to determine the voltage drop across the capacitance

$$U_C = \frac{U_{in} \cdot d}{4 \cdot \pi \cdot R \cdot f \cdot S \cdot \varepsilon \cdot \varepsilon_0 + d} \tag{3}$$

Using simple transformations, we obtain the expression for lubricant film thickness from (3)

$$d = \frac{4 \cdot \pi \cdot U_C \cdot R \cdot f \cdot S \cdot \varepsilon \cdot \varepsilon_0}{U_{in} - U_C} \tag{4}$$

A micro-controller with integrated 12-bit multichannel ADC is used to measure U_{in} and U_C. After a full measurement cycle, filtering and averaging of the measured values using the floating-point unit (FPU) integrated in micro-controller, the lubricant film thickness is calculated according to the formula (4).

However, such a measurement approach is prone to the above-stated significant drawback: a significant dispersion in readings over time and small ranges of measurement of contacting film. This required an improvement in the design of the device.

Which was achieved by including into the device's electrical circuit a digital controlled-amplitude generator implemented on a twisted-ring counter.

Functionally, the instrument consists of a measuring transducer and a power supply unit (figure 2).

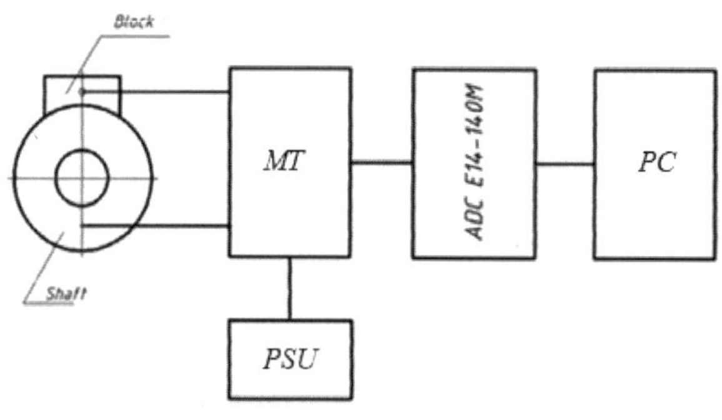

Figure 2. Device for measuring the lubricant film thickness, a block diagram: *MT* – measuring transducer; *PSU* – power supply unit; *PC* – personal computer.

The measured capacitance is connected to one of the bridge arms (see figure 3); sinusoidal voltage is supplied from the generator "*G*" to the bridge's upper point, and then to the synchronous detector "*SD*". From the bridge, the voltage is fed to the repeaters "*R*", from where it is supplied to the differential amplifier "*DA*" and then to the upper input of the "*SD*".

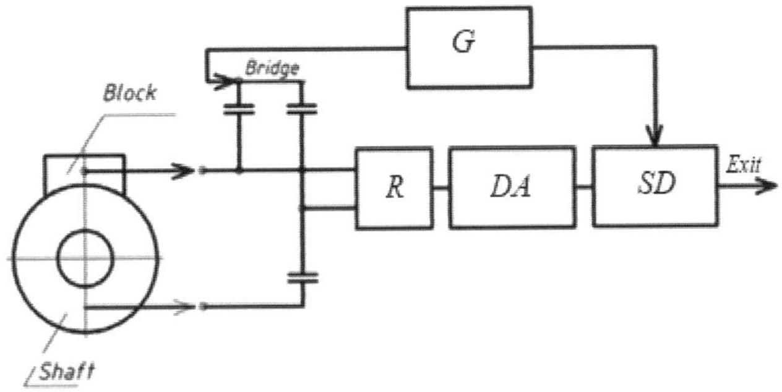

Figure 3. Measuring transducer, a functional diagram: *G* – generator; *DA* – diode amplifier; *SD* – synchronous detector; *R* – repeaters.

At the output of the synchronous detector, the pulsating voltage is converted into a constant voltage using a filter (figure 4).

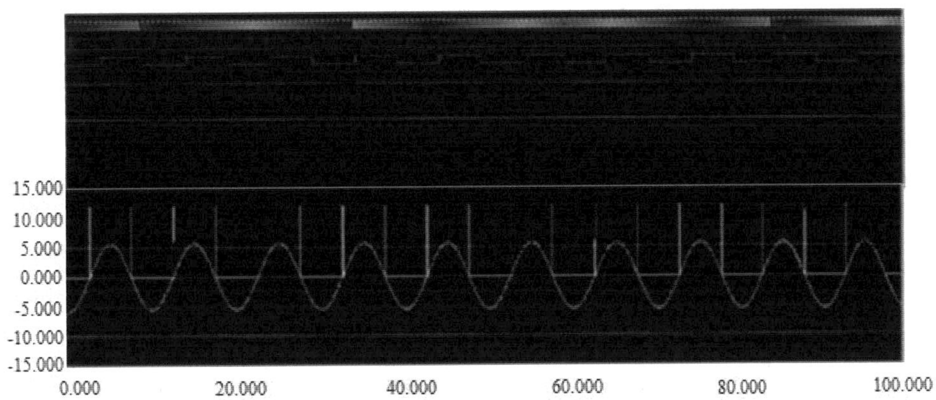

Figure 4. Synchronous detector operation

The device was tested on a modernized friction machine II 5018 [9].

The procedure for preparing for measurements includes the following steps:

- isolation of the roller (or the holder of the block), the loading mechanism, from the friction machine by placing a ring of dielectric material on the upper shaft;
- installation of current collectors of the upper and lower shafts, to improve the precision of measurements and to gauge the electrical signal from the upper and lower shafts;
- connecting to the ADC the device for measuring the lubricant film thickness (figure 2);
- connecting the ADC to a PC.

The new device was calibrated using polyethylene films of various thickness, with dielectric permeability close to that of lubricating oil.

The films were installed alternately between the test specimens, under a load that had insignificant impact on their deformation.

In this case, the dispersion in instrument readings stayed within ±5%.

The calibration curve is shown in figure 5.

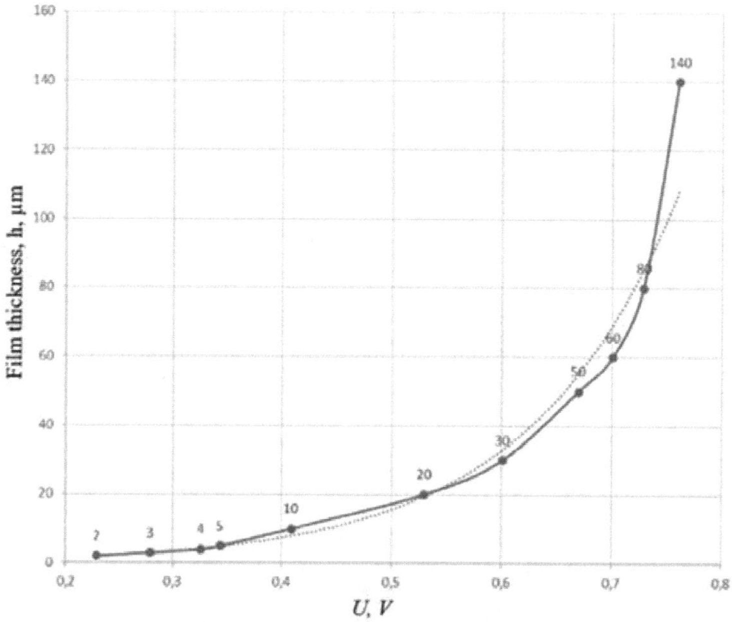

Figure 5. Calibration curve of the device for measuring the lubricant film thickness.

4. Results

Experimental studies were carried out according to the "roller-block" scheme (the roller is made of "45 GOST 1050-2013" steel, the block is made of "BrAZh-9-4 GOST 18175-78" material). The block had a different diameter clearance, 15 and 30 μm [10].

Before the tests, the friction pairs had additional run-in. The roughness of the surfaces was determined using a surface analyser.

The results of measuring the lubricant film thickness were displayed on a computer screen as oscillograph recording. (figure 6, (a), (b)).

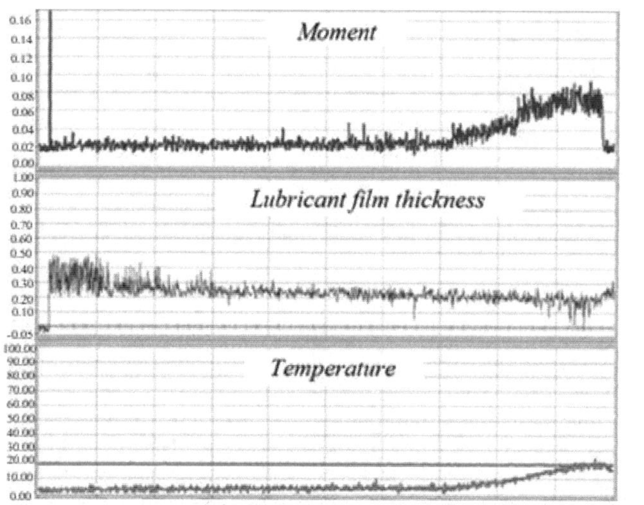

(a)

Figure 6. Oscillograph recording of lubricant film thickness in "roller-block" friction pair (steel "45" – "BrAZh 9-4"), V = 0.5 ms^{-1}, lubricant I-40A (flow rate 0.03125 min^{-1}) with diameter clearance:
a) 15 μm (Ra = 0.3 for the roller, = 0.8 for the block)
b) 30 μm (Ra = 0.3 for the roller, 0.7 for the block).

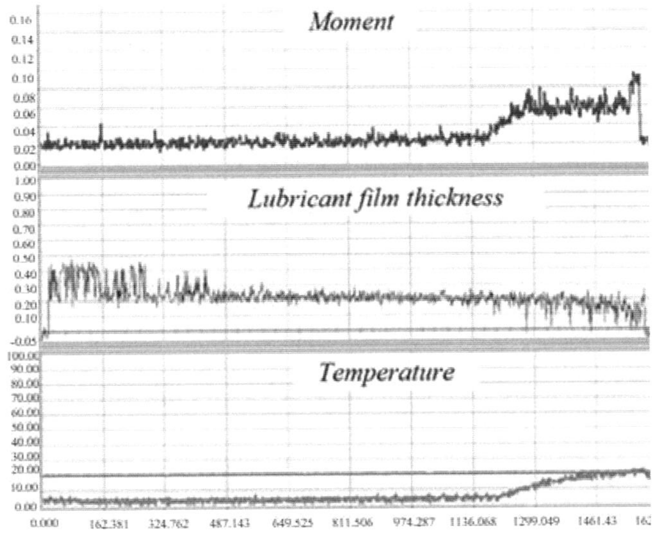

(b)

To reduce labour intensity and increase the data flow during the experiment, the test chamber was modernised.

Liquid lubricant "I-40A GOST 20799-88" was circulated into the friction zone using a proprietary-design system [9].

The temperature in the contact zone was determined using calibrated thermocouples.

The test results are presented in the table 1 and in figure 6.

Table 1. Results of measuring the lubricant film thickness in the contact of "roller-block" friction pair (steel "45"- "BrAZh 9-4").

Diameter clearance, δ, μm	Specific contact pressure, p, MPa	Average value of the instrument readings for measuring the lubricant film thickness, U, V	Lubricant film thickness, h, μm
15	2.5	0.270	2.92
	5	0.250	2.52
	7.5	0.240	2.34
	10	0.234	2.24
	12.5	0.229	2.16
	15	0.223	2.07
	20	0.210	1.88
	25	0.180	1.51
30	2.5	0.282	3.19
	5	0.257	2.65
	7.5	0.238	2.31
	10	0.230	2.17
	12.5	0.218	1.99
	15	0.212	1.91
	20	0.172	1.42
	25	0.110	1.04

5. Discussion

Experimental studies have shown that the proposed approach for measuring the lubricant film thickness allows obtaining stable readings under identical test conditions, with a dispersion of ± 5% and minimum thickness of a lubricating film from 1 μm (see the table).

It was found that at low roller rotation speeds (V = 0.5 ms^{-1}) and high specific pressures, the lubricant film thickness is higher in a friction pair with 15 μm diameter clearance compared to 30 μm diameter clearance.

6. Summary

1 An original device for measuring the lubricant film thickness has been developed, tested and introduced into the practice of tribological research. Its principle of operation is based on the capacitive method using a generator implemented on a twisted-ring counter.

2 The paper presents the method of calibration of a device for measuring the lubricant film thickness, based on the use of films of various thickness, with dielectric permeability close to that of lubricating oil.

3 It was found that the proposed device ensures obtaining stable results with a dispersion of ±5%, allowing for measurements of the thickness of contacting film starting from 1 μm.

References

[1] Zubkov N N, Vasiliev *et al* 2014 Modernization of Amsler-type friction machine *J. Friction and lubrication in machines and mechanisms* **9** pp 33-37

[2] Al-Saadi D A Yu, Pershin V F *et al* 2017 Modification of graphite lubricants with graphene nanostructures *J.Friction and Wear* V **385** pp 418-22

[3] Kulakov O I and Albagachiev A Yu 2019 Modernization of II 5018 friction machine *Proc. VI Int. Sc. Conf. Fundamental Research and Innovative Technologies in Mechanical Engineering* (Moscow: Federal State Budgetary Institution of Science, Blagonravov's Institute of Mechanical Engineering, Russian Academy of Sciences) pp 244-45

[4] Levchenko V A, Buyanovskiy I A *et al* 2019 Green tribology: orientation properties of carbon diamond-like coatings of tribological units in lubricating media (review) *J. of Applied Chemistry* **12** pp 1499-1513

[5] Reshchikov V F 1975 *Friction and wear of heavily loaded gears* (Moscow: Mechanical Engineering) p 232

[6] Snegovsky F P 1969 *Sliding bearings for heavy machines* (Moscow: Mechanical Engineering 1969) p 220

[7] Kodnir D S 1976 *Contact hydrodynamics of lubrication of machine parts* (Moscow: Mechanical Engineering) p 304

[8] Drozdov Yu N, Pavlov V G *et al* 1986 *Friction and wear in extreme conditions: Handbook* (Moscow: Mashinostroenie) p 223

[9] Darovskoy G V, Burakova M A, Polyakov V N and Krotov V N 2017 *Investigation of the frictional properties of high traction oils* (Rostov-on-Don: RGUPS) p 88

[10] Darovskoy G V, Krotov V N, Polyakov V N and Ezupova M N 2019 Modeling the hydrodynamic friction mode on Amsler-type friction machines *J.Friction and Wear* V **403** pp 284-90

Study of the effects of copper undecylate on the tribological properties of refined waste oil

S V Fedosov[1], V V Terent'ev[2], A V Markelov[3], Yu P Osadchiy[3] and A V Sokolov[4]

[1] National Research Moscow State University of Civil Engineering, Moscow, Yaroslavl Highway, 26, RF
[2] Ivanovo State Agricultural Academy, Ivanovo, 45 Soviet Street, RF
[3] Ivanovo State Polytechnic University, Ivanovo, 21 Sheremetev Avenue, RF
[4] Yaroslavl Technical State University, Yaroslavl, 88 Moscow Avenue, RF

aleksandr203.37@mail.ru

Abstract. The results of experimental studies of the use of refined spent oil as a base for manufacturing of lubricants with additives in the form of copper undocylate are presented. The effectiveness of cleaning spent oil by baromembran methods has been shown. The introduction of copper undecylate into the refined spent oil can reduce the friction factor by 1.7 times, wear by 28%.

1. Introduction

The effectiveness of the use of lubricants in various tribosystems is determined not only by the characteristics of the base oil, but also by the effectiveness of the functional additives and fillers used in it. In the process of work in the friction units, the oils used are contaminated, oxidized, which leads to deterioration in their performance characteristics and, ultimately, to loss of performance. Therefore, during periodic maintenance of equipment, the oils are replaced with fresh ones. This raises problems associated with the disposal of contaminated and discarded oil. One of the promising ways to improve the efficiency of the use of lubricating oils is their purification and restoration of operational characteristics to the extent required for the effective operation of tribo-coupling. Currently, various schemes for the purification of contaminated oils have been developed, based on the use of both physical and chemical methods [1-9]. Among the variety of currently existing methods of deep purification of oils, the method of baromembrane filtration can be noted. Depending on the pore size of the membranes, both micro- and ultrafiltration can be carried out. The use of various adsorbents and coagulants can improve the cleaning efficiency.

In the future, refined oils can be used as abasis of lubricant composition, adding a complex of functional additives to them, depending on the conditions for further use in tribo-conjugations. Previous studies [10-11] point to the promising use of various liquid crystal compounds and their mixtures as antifriction and anti-wear additives to lubricants.

2. Goals and objectives of the study

The aim of the study is to develop a method for recycling waste oils by purifying them and introducing effective anti-wear and antifriction additives to obtain lubricants.

3. Materials and methods

A mixture of used engine and transmission oils is used as a starting material. They are cleaned in a multi-stage unit. During the cleaning process, waste oils are subjected to coagulation, separation and ultrafiltration using polymer or ceramic membrane filters. This installation allows to reduce the effect of dispersant additives and to separate mechanical impurities, water and oil aging products.

The design of the installation is shown in figure 1. The used engine oil flows from the reservoir with the initial solution 1 into the reactor 3 by gravity. With the help of a metering pump 12, a coagulant is fed into the reactor from the reservoir 2. The reactor is mixing the used engine oil and coagulant for 20 - 30 minutes. To maintain optimal temperature conditions during separation into the reactor, an electric heating device 22 with automatic temperature control is installed.

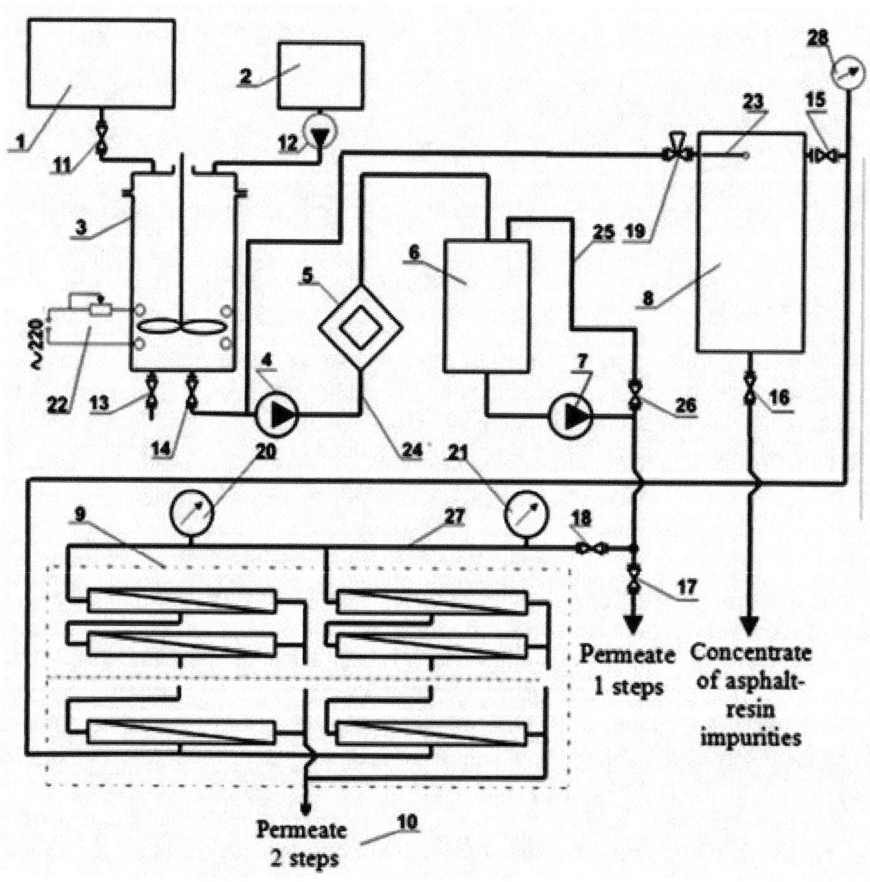

Figure 1. Proposed installation scheme for a two-stage recovery of used engine oil: 1 - reservoir with original waste oil; 2 - reservoir with coagulant; 3 - reactor; 4 - pump of the first stage of cleaning; 5 - separator; 6 - intermediate tank; 7 - pump of the second cleaning stage; 8 - reservoir for the concentration of asphalt-resinous impurities; 9 - blocks of membrane tubular modules; 10 - tank with purified waste oil;11, 13, 14, 15, 16, 17, 18, 26 - throttle valves; 12 - dosing pump; 19 - solenoid valve; 20, 21, 28 - manometers; 22 - thermostat; 23 - level sensor; 24 - pressure line of the 1st stage; 25 - bypass hydraulic line; 27 - pressure line of the 2nd stage.

Then the waste oil treated with the coagulant is fed by the gear pump 4 to the separator 5, where pre-treatment takes place for 15 - 20 minutes. Waste engine oil, pre-purified from mechanical impurities, water and partially from asphalt-resinous compounds, enters the intermediate tank 6.

Thus, the first stage of filtration ends.

Further, the resulting solution can be used as a finished product or be subjected to deeper cleaning. For this, the pump 7 supplies the solution to the parallel-series-connected blocks of membrane tubular modules 9.

Technological modes of the separation process, the pressure drop and the flow rate of the solution through the membrane modules are carried out using throttle valves 15, 18, 26. After the separation process, the concentrate enters the tank 8, the filtrate (permeate) is discharged into the storage tank with purified oil 10.

To maintain the required amount of solution in the tank 8, a level sensor 23 is provided. The piping system, shut-off valves and the solenoid valve 19 provide for continuous filling of the tank with the original used engine oil.

The proposed model does not contain complex mechanical systems, which increases its reliability, and the use of pre-treatment in front of the membrane apparatus allows for improved quality of the permeate.

The analysis of the obtained samples of refined oil was carried out in the accredited laboratory "Ivanovo chemical-technical laboratory" of the Northern Railway of the branch of JSC "Russian Railways".

Copper undecylate was added to the purified spent oil to study the possibility of improving tribological characteristics. This compound was selected on the basis of earlier studies [10, 11], where it was shown that copper carboxylates, which are discotic mesogens, under certain conditions in mixtures with lubricants also exhibit liquid crystalline properties, while copper undecylate exhibits them already at room temperature.

The study of tribological characteristics was carried out on an SMT-1 friction machine according to the "stationary ball - rotating roller" scheme [19, 20]. The test roller was lubricated by immersing it in an oil bath. In the process of research, the coefficient of friction, ball wear, and the temperature of the oil in the bath were determined. The load on the samples was 200 N.

4. Results and discussion

Table 1 presents the characteristics of the oil before and after cleaning.

Table 1. Oil characteristics before and after the refining process.

Indicator name	Indicator value	
	before cleaning	after cleaning
Kinematic viscosity of oil at 313 K, cSt	54.84	26.61
Kinematic viscosity of oil at 323 K, cSt	37.74	19
Kinematic viscosity of oil at 373 K, cSt	9.77	5.43
Density, kg·m^{-3}	861	847
Oil purity, %	97.7	99.6
The dielectric constant	2.6293	2.4362
Base number, mg KOH g^{-1}	6.4	4.0

As can be seen from table 1, the use of ceramic membranes allows deep cleaning of the oil, while most of the additives are removed from the oil. This is evidenced by a decrease in kinematic viscosity by 1.8-2 times, density and base number by 1.6 times. Water and mechanical impurities are completely removed from the oil. However, no significant oil clarification occurs during the cleaning process.

The results of determining the coefficient of friction are shown in figure 2.

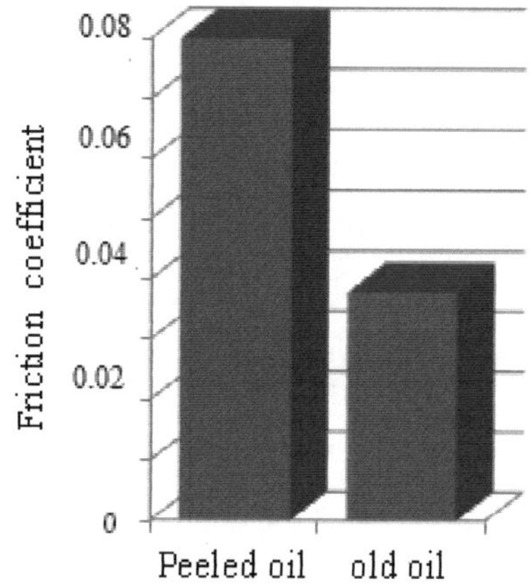

Figure 2. Results of determining the coefficient of friction.

As can be seen, the introduction of copper undecylate into the oil leads to a 1.7-fold decrease in the friction coefficient.

Thus, copper undecylate can effectively improve the antifriction characteristics of used oil.

To increase the resource and durability of friction pair, it is required to reduce their wear rate.

Improving the antifriction characteristics of the lubricant by introducing mesogenic copper undecylate makes it possible to improve the operating mode of the friction pair and to reduce the wear rate of the friction pair elements.

Figure 3 shows the results of determining the wear of a stationary sample.

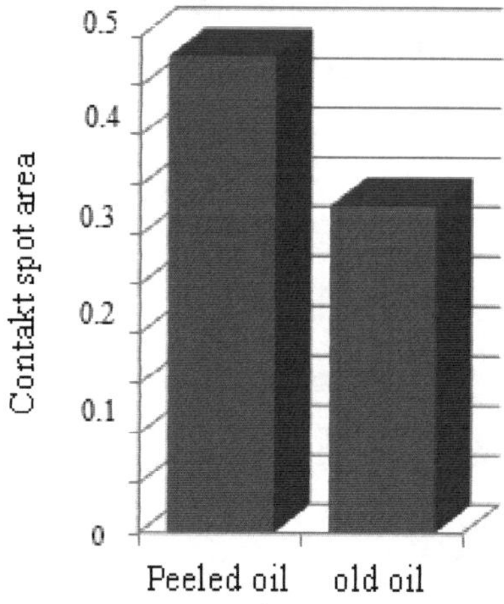

Figure 3. The results of determining the wear of a stationary sample.

The addition of copper undecylate to the purified oil leads to a decrease in the wear of the stationary sample (ball) by 28%, which is comparable with the results obtained earlier [10, 11]. At the same time, the increase in oil temperature during the test in both cases was 293 K.

IOP Conf. Series: Materials Science and Engineering **996** (2020) 012011 doi:10.1088/1757-899X/996/1/012011

We associate the improvement in the tribological characteristics of the refined oil with the introduction of the carboxylate with the transition of copper undecylate into the mesophase in the friction zone and with its particular orientation on the friction surface.

5. Conclusion

Thus, copper undecylate is a fairly effective antifriction and anti-wear additive to hydrocarbon-based lubricants and its use in the future will increase the resource of tribo-couplings and the durability of the machine as a whole.

This study also shows that waste oils, after cleaning and restoration of tribological properties, can be used as lubricants in various mechanisms, which makes it possible to solve problems with their disposal.

References

[1] Svitzov A A 2007 *Introduction to membrane technology* (Moscow: DeLi print) p 280
[2] Ignatovich E 2007 *Chemical engineering. Processes and apparatus* (Moscow: Technosphere) p 656
[3] Rodionov A I, Kuznetsov Yu P and Soloviev G S 2005 *Protection of the biosphere from Industrial emissions. Basics of designing technological processes* (Moscow: Chemistry, KolosS) p 392
[4] Dytnersky Yu I 1995 *Processes and devices of chemical technology. Textbook for universities. Theoretical foundations of chemical technology processes. Hydromechanical and thermal processes and devices* (Moscow: Chemistry) p 400
[5] Lacey R E and Loeb S 1976 *Technological processes using membranes* (Moscow: Mir) p 370
[6] Hvang S-T and Kammermejer K 1981 *Membrane in separation* (New York, London, Sydney, Toronto: A wiley-interscience publicaton. John Wiley & Sons) p 464
[7] EinsteinV G, Zakharov M K, Nosov G A and others 2000 *General course on processes and devices of chemical technology* (Moscow: Chemistry) p 1760
[8] Fedosov S V, Maslennikov V A, Osadchiy Yu P and Markelov A V 2020 Pressure loss along the channel of a tubular membrane during the ultrafiltration of liquid media (*Theoretical Foundations of Chemical Engineering*) chapter 2 pp 380-87
[9] Fedosov S V, Osadchiy Yu P and Markelov A V 2020 Modeling the ultrafiltration process taking into account the formation of sediment on the membrane surface (*Membranes and membrane technologie*) chapter 3 pp 177-89
[10] Terentyev V V, Akopova O B and Telegin I A 2015 Influence of additives from mixtures of copper carboxylates on the tribological characteristics of greases (*Liquid crystals and their practical use*) chapter 4 pp 96-101
[11] Terentyev V V, Akopova O B and Telegin I A 2017 Influence of mesogenic copper behenate additive on rheological and tribological characteristics of greases (*Liquid crystals and their practical use*) chapter 1 pp 93-100
[12] Zinoviev E V, Levin A L, Borodulin M M and Chichinadze A V 1980 *Polymers in friction units of machines and devices: Handbook* (Moscow: Mechanical engineering) p 208
[13] Melnik S V and Goloshchapov G A 2016 Application of modified greases in conditions of abrasive wear of friction pairs of tracked running devices (*Bulletin of modern science*) chapter 9 pp 56-61
[14] Rebinder P A and Fuchs G I 1973 *Advances in Colloid Chemistry* (Moscow: Nauka) p 359
[15] Akhmatov A S 1963 *Molecular Physics of Boundary Friction* (Moscow: Fizmatgiz) p 472
[16] Fuchs G I *Research in the physical chemistry of contact interactions* (Ufa, Bashkir book publishing house) p 221
[17] Kadkin O N, Domracheva N Ye and Ovchinnikov I V 2015 Metal mesogens: a retrospective and perspective view (*Liquid crystals and their practical use*) chapter 4 pp 21-39
[18] Chandrasekhar S 1980 *Liquid Crystals* (Moscow: Mir) p 345

[19] Gaidar S M 2015 *Planning and analysis of an experiment: textbook* (Moscow: publishing house of FSBNO "Rosinformagroteh") p 548
[20] Pruzhansky L Yu 1978 *Investigation of test methods for wear* (Moscow: Nauka) p 116

Acknowledgments
This work was financially supported by the FSBO "Fund for Assistance to the Development of Small Forms of Enterprises in the Scientific and Technical Sphere" for 2020–2021 (contract No. 3642GS1 / 60673 dated 06.08.2020).

IOP Conf. Series: Materials Science and Engineering **996** (2020) 012012 doi:10.1088/1757-899X/996/1/012012

The use of boehmite in tribology

A V Fedotov[1], A V Gvozdev[2], A V Dunaev[1] and I A Pekhalsky [1]

[1] Federal scientific Agroengineering center VIM, Moscow, Russia
[2] GBOU VO Ivanovo state agricultural Academy, Ivanovo, Russia

fedotov48@list.ru

Abstract. The analysis of the results of the use of nanostructure boehmite as a break-in and prophylactic tribological preparations is carried out. The mechanism of boehmite action during the running-in of a diesel engine and a drilling rig gearbox is considered. A boehmite-based product reduces break-in time, reduces the friction coefficient and wear of tribological couplings. A scheme of the effect of boehmite on increasing the life of tribological compounds is developed. Justified its use as a repair and recovery tribological product.

1. Introduction

Losses of financial resources from friction and wear in developed countries reach 4-5% of national income, and overcoming friction resistance absorbs 20-25% of the energy generated per year around the world. For the full cycle of operation of machines (in transport, energy, engineering, etc.), operating costs, repair costs and spare parts are several times higher than the cost of manufacturing new equipment [1-3]. Therefore, research and the search for new tribological compositions that can reduce friction, wear, and operating costs continue. And nanostructure boehmite may be one of such materials [4].

2. The goals and objectives of the study

Evaluation of the effectiveness of boehmite in tribotechnics: when running in engines and transmission units, as a preventive measure when introduced into oils and greases.

3. Materials and research methods

We used boehmite powder obtained by hydrothermal oxidation of aluminum powder [5]. A distinctive feature of this boehmite is its high degree of homogeneity, the stability of the composition and structure. According to x-ray structural and petrographic analyzes, the material is well crystallized and consists entirely of aluminum monohydroxide (boehmite) AlOOH. (figure 1 (a)). The coherent scattering region characterizing the crystal size of the initial powder is 20–40 nm for different batches of the powder and does not exceed 80 nm. The boehmite of hydrothermal synthesis can be characterized as a nanostructured material. Losses on ignition correspond to stoichiometric 15 wt. % Using a scanning microscope, it was possible to observe both individual particles and particle aggregates predominantly 2-3 microns in size, up to a maximum of 10 microns (figure 1 (b)). Individual particles of irregular shape, consisting of smaller nanocrystallites, are combined into large aggregates.

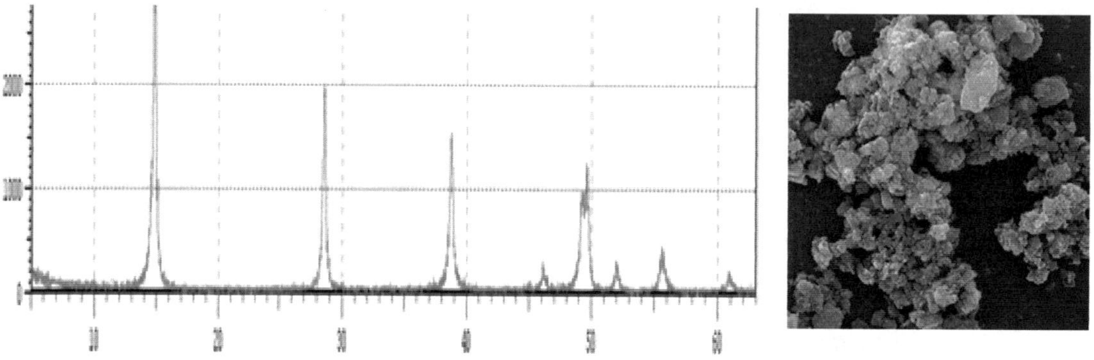

(a) (b)

Figure 1. X-ray powder of boehmite (a) and a scan of the aggregates of its particles (b).

As you know, a common drawback of nanoparticles is the tendency to aggregation, which eliminates their valuable properties. Therefore, nanostructure boehmite powder or its composition with other mineral components was mixed with a surfactant, introduced into oil, and sonicated. This ensured the disaggregation of powder particles, the aggregative and sedimentation stability of the suspension.

Comparative tests of 12 tribotechnical materials, including those containing serpentine additives, were carried out according to the "roller-block" scheme (block — cast iron of cylinder liner D-240 diesel, roller — steel 40Cr) on a friction machine 2070 SMT-1M with a rotational character of movement test samples. The coefficient of friction and the temperature in the mating zone of the parts were determined.

Comparative wear tests were carried out on a modernized 77MT-1M friction machine with a smaller sample (a fragment of a piston ring) reciprocating along a larger one (a fragment of a cylinder liner D-240), at a load of 300 N and a sliding speed of 3.2 ms^{-1}. Accelerated tests were carried out in conditions of physical modeling of the real friction process (identical materials and kinematics of the translational movement of the piston along the surface of the cylinder). Pressure was the accelerating factor.

The effect of boehmite on the run-in was studied on a D-243 diesel engine at the upgraded break-in brake stand KI-5540-GOSNITI and a drilling rig gearbox. In the second case, kaolin and talc were introduced with boehmite. We used kaolin of the Prosyanovskoye deposit according to TU 421-533-2001, talc of the Shabrovskoye deposit of TMK-28 grade. Run-in of aggregates was carried out according to standard methods

The surface roughness was checked before and after running-in by a Surtronic-ZR profilometer from Taylor-Hobson (England). The wear of the mating parts of a diesel engine at various stages of running-in was evaluated by the mass fraction of solids in oil.

4. Results and its discussion

On a friction machine 2070 SMT-1M tribostructure with boehmite showed the lowest (0.045) coefficient of friction in a wide range of loads (150 - 600 N, up to 150 MPa). He increased the load capacity of the tribocouple in comparison with pure oil almost 2 times. At constant load the temperature did not exceed 60 °C, and with pure oil it rose to 90 - 100°C.

On the friction machine 77MT-1M, the composition containing boehmite showed the smallest wear (wear spot length decreased from 2.45 mm for pure oil to 1.35 mm, i.e., 1.81 times). The wear rate calculated as the ratio of the wear value (spot length) in m to the friction path in m was $1.06 \cdot 10^{-7}$ for pure oil, and the inverse value, wear resistance, was approximately $0.94 \cdot 10^{7}$. For tribostructure with boehmite, respectively $0.60 \cdot 10^{-7}$ and $1.70 \cdot 10^{7}$. The cylinder-piston group of the car spot length corresponds to the 11-12 wear resistance class or 10^{11} - 10^{12}, which corresponds to the wear rate of

10^{-11} - 10^{-12}. The wear value obtained in the experiments is greater by 4-5 orders of magnitude. This manifests the effect of accelerated testing under accepted conditions on a friction machine 77MT-1M.

In addition to tests on a friction machine 77MT-1M, we studied the effect of boehmite on wear on the four ball friction machine (FBFM) and on the MTU friction machine. Boehmite was introduced into motor, industrial, rapeseed oil and solid oil. Depreciation was estimated by the size of the wear spot and weight loss of the mating parts. Tests have shown that the addition of boehmite reduces the size of the wear spot at (FBFM) to 34%, the weight loss of parts (MTU machine) decreased by 2 and 6 times, respectively, for grease (solid oil) and I I20 oil.

Run-in tests have shown that introducing boehmite powder and surfactant into M-10G_{2K} engine oil reduces the time to reach maximum compression (P) by 2 times, ensures quick achievement of rated power values (W), reduces fuel consumption (G) and oil temperature (T), the roughness of the bearing shells and piston rings decreases by 1.4 and 1.7 times, respectively (table 1). In general, the addition of boehmite accelerates and improves the quality of running-in, reduces running-in wear by up to 2.5 times, which increases the operational life of a diesel engine.

Table 1. The results of running-in diesel D - 240 with (1) and without boehmite (2).

Time,	*P*, MPa		*W*, kW		*G, g / h.p.*		*T*, °C	
min	1	2	1	2	1	2	1	2
40	3.0	2.93					30	40
80		3.0					60	80
90			52.5	45.0	257	270		

Lapping additive on the basis of boehmite showed itself well in terms of running-in of engine parts of grade 10-11 of cleanliness and hardness of 52-56 HRC. The teeth of the gear wheels (steel type 18HGT) and rolling bearings (steel type SHX-15) have a hardness of 61-63 HRC or 179 - 207 MPa with a cleanliness class of 7-8 with a roughness of Rz 3.2-6.0 microns. The nanoscale of boehmite particles (crystal size less than 100 nm, aggregates less than 1 μm) leads to their indentation into the hollows of the roughness of the rough surfaces of the teeth and, obviously, therefore, the abrasiveness of boehmite powder in the transmission units was weak. In addition, during friction at elevated temperatures in the contact of mates, disintegration of boehmite particles and its decomposition with the release of water and the formation of aluminum oxides can occur. In this case, the grinding of the starting aggregates and individual particles occurs, which further reduces their size [6].

Also the additive, contained of nanostructure boehmite, surfactant, talc and kaolin and which is used in running transmission units, was developed. Its tests were carried out on the gearbox of a mobile drilling rig mounted on the chassis of a KamAZ car. A feature of the gearbox is the presence of high hard gears and bearing parts, the absence of non-ferrous metals and soft alloys.

Compared with the run-in on a standard gear oil TM-3, a reduction in the duration of the run-in of the gearbox from 8 hours (working shift) to 3.0-3.2 hours, i.e. 2.5 times, was obtained; decrease in oil temperature from 75 to 61°C; 2.5 times increase in the running-in surface area. At the same time, after draining the break-in oil and refueling, the noise and the temperature of heating the parts decreased, and an increase in the resource of the critical unit of the drilling rig is predicted.

The positive effect of kaolin and talc additives is due to their physicomechanical properties and particle sizes (units μm), suitable for running gears, comparable with the surface roughness of gear parts. Kaolin and talc are thermally more stable and do not decompose at temperatures of 400-600°C, like boehmite, with the formation of smaller particles. Particles of talc and kaolin abrade microroughnesses in the surfaces of parts, increasing the area of the contact contact surface, reducing unit loads and, thus, preventing the formation of a scuff. Once in the friction zone, leveling the surface, these particles can adsorb resins and oxidation products of the lubricating medium. Kaolin with larger particles is more actively involved in abrasive grinding, opens juvenile surfaces and promotes a more active effect of surfactants according to the acad effect P.A. Rebinder. Talc, as a plastic material, serves as an intermediate material between kaolin particles and, together with boehmite, promotes a more uniform distribution of kaolin particles in the lapping composition and in the running oil. Thus, the running-in efficiency depends not only on the physicomechanical properties of the components of the additive, but also on the ratio of the particle size of the components and the surface roughness of the parts being worked on. For details of greater roughness and hardness, compositions of boehmite with minerals are effective larger components.

Taking into account the provisions of tribology, the role of boehmite in tribotechnology can be reduced to the following:

- abrasive wear and grinding of areas of increased roughness of friction joints, increasing the purity of their surfaces, reducing the mechanical component of the coefficient of friction;
- cleaning of friction surfaces from deposits, oxide films and defective structures, which ensures access of tribomedia substances to juvenile surfaces of metals and accelerates the formation of antifriction coatings;
- some subsequent adsorption on the surfaces of boehmite particles of resinous substances creating a "third body" in the tribopair, which also reduces the coefficient of friction;
- dispersion and decomposition of boehmite (with the release of water), which reduces the overheating of contact surfaces;
- lamination and hydration shells on the surface of the particles also contribute to a decrease in the coefficient of friction [7];
- a catalytic effect on chemism in the friction zone is possible;
- nanostructured ceramicization of local defects with the formation of a layer of high heat resistance, wear resistance and high electrical resistance.

Each of the listed properties can manifest itself under different friction conditions, at different stages of the transformation of boehmite particles: at the beginning, some properties play an important role, and later on others. With this in mind, a scheme has been developed that explains the increase in the resource of friction units by reducing their wear rate (figure 2), running-in wear, and shortening the break-in time.

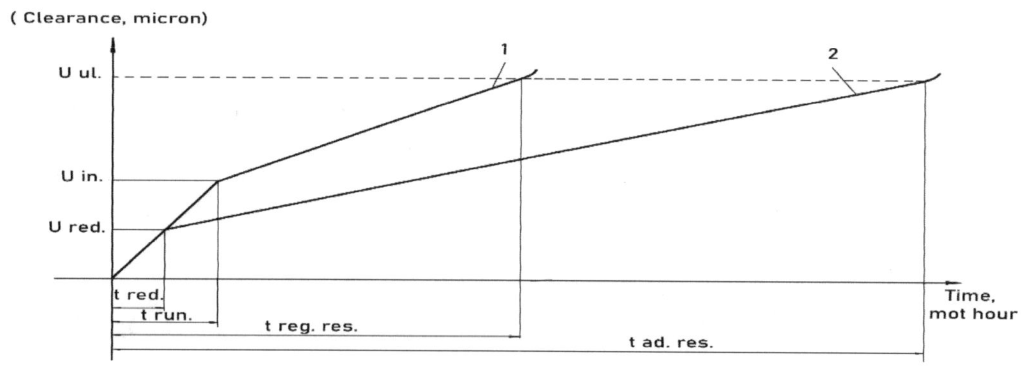

Figure 2. Wear of the interfaces before (1) and after (2) the introduction of tribostructure. *U red.* - reduced initial clearance (wear) with the introduction of tribostructure; *U in.* - the initial clearance of the interface during run-in; *U ul.* - ultimate clearance; *t red.* -reduced run-in time of the unit with the introduction of tribostructure; *t run.*– run -in time of the unit; *t reg. res.*- regulatory resource; *t ad. res.* - an additional resource due to the use of tribostructure.

It is known that with significant wear and tear to extend the service life of the units can provide repair and restoration compounds. [2, 3]. Especially effective of them are serpentine. In this regard, it is interesting to compare the physicochemical properties of serpentine minerals and boehmite (table 2).

Table 2. Some properties of serpentine and boehmite powders.

Properties	Serpentine	Boehmite
The water content, %	13	15
Decomposition temperature, °C	650 - 750	500 - 600
Mohs hardness	2.5-4.0	3.5
Compounds with FeO, Fe_2O_3	$FeO\cdot Al_2O_3$,$Fe_2O_3\cdot Al_2O_3$	$2FeO\cdot SiO_2$, $MgO\cdot Fe_2O_3$
Energy density, $J\cdot cm^{-3}$	82.78	104.31
Catalytic effect	+	+

Mohs hardness of serpentine minerals 2.5-4.0, boehmite - 3.5. The total amount of constitutional water is 13% for serpentine, 15% for boehmite. The final decomposition products of serpentine are members of a continuous isomorphic series of olivines containing iron oxide and silica, and in boehmite – alumina.

The end product of boehmite decomposition, alumina, can also form chemical compounds with iron oxides, as well as serpentine decomposition products, including solid solutions. The Fe_2O_3 – Al_2O_3 system was studied by various authors [8, 9]. The crystalline phases present in the system are: hematite with corundum in solid solution, corundum with hematite in solid solution, phase 1: 1 = $Fe_2O_3 \cdot Al_2O_3$ (solid solution), spinel phase, which is mainly a solid solution between magnetite ($FeO \cdot Fe_2O_3$) and hercinitis ($FeO \cdot Al_2O_3$) with some excess of Fe_2O_3 or Al_2O_3, or both at the same time. The formation of a solid solution of $Fe_2O_3 \cdot Al_2O_3$ is promoted by the same structure and close parameters of the unit cell of oxides. Thus, both the decomposition products of serpentine and aluminum oxides, which are the decomposition products of boehmite, can enter into chemical interaction with iron oxides always present on the surface of steel parts with the formation of new phases.

The energy density of boehmite is higher than that of serpentine, which can contribute to the formation of repair and restoration tribo-coating [10].

The use of serpentine tribo compounds leads to the formation of tribocover containing mainly carbon [11]. Under conditions of friction, oil is pyrolyzed with carbon evolution, and serpentine decomposition products can be catalysts for this process. But it is also known that boehmite is widely used for the manufacture of catalyst supports and catalysts themselves. Therefore, we can assume the formation of carbon tribocovering, as a result of the catalytic action of boehmite, as boehmite in many physicochemical properties (water content, dehydration temperature, hardness, energy density, ability to form chemical compounds and solid solutions with iron oxides, catalytic effect) has properties similar to serpentine. The above suggests that boehmite can be not only run-in, but also repair-restoration composition, which of course requires additional tribological confirmation.

Alpha-alumina or corundum is the most stable form of alumina - the last phase in the process of boehmite transformations. Corundum is chemically resistant to many reagents and melts, has insulating properties, high hardness (9 on the Mohs scale), and is practically not hygroscopic. High hardness of corundum is not a negative factor. More important is the particle size. For example, many tribological compositions include nanodiamond particles (KARAT-5, KARAT-M, Formula «AB», Renom Engine Lubrifilm, Diamond Run In, Nanodiamond Green Run, etc.) [11]. These compounds have proven themselves as preventive, and KARAT-5 and KARAT-M - and as repair and restoration. In running-in, they reduce running-in wear, accelerate running-in, and realize equilibrium surface roughness. As a result, wear resistance is increased, mechanical losses are reduced, and the operational qualities of the units are improved. The "nanodiamonds" that make up the structure of the oil film, increasing its dynamic strength, form new friction surfaces, and reduce (especially at high loads and lubricant deficiency) friction and wear.

5. Conclusion

The introduction of nanostructure boehmite and surfactant into the oil up to 2 times accelerates and improves the quality of running-in of internal combustion engines, ensures the rapid achievement of passport values of power and fuel consumption. Running-in composition for running in gearboxes containing boehmite, kaolin, talc and surfactant provides a 2.5-fold reduction in running time and temperature, an increase in the running-in surface area. The running-in efficiency depends not only on the physicomechanical properties of the components of the additive, but also on the ratio of the particle size of the components and the surface roughness of the worked-in parts. Compositions based on nanostructure boehmite can be promising running-in and prophylactic tribological compositions that provide savings in material and energy costs.

References

[1] Chichinadze A V, Berliner E M, Brown E D et al 2003 *Friction, wear and lubrication* ed A V Chichinadze (Moscow: Mechanical Engineering) p 576
[2] Garkunov D N 2001 *Tribotechnology (wear and tear)* (Moscow: Publishing MCXA) p 616
[3] Chernoivanov V I 1995 *Restoration of machine parts* (Moscow: GOSNITI) p 278
[4] Chernoivanov V I, Sergeev N N, Dunaev A V and Fedotov A V 2013 *Additive composition for running-in oil for running in an internal combustion engine and running-in oil* RF Patent No 2472848
[5] Mazalov Yu A, Bersh A V and Ivanov Yu L 2006 *A method of producing hydroxides or oxides of aluminum and hydrogen and a device for its implementation* RF Patent 2278077
[6] Bersh A V, Mazalov D Yu, Soloviev R Yu, Sudnik L V and Fedotov A V 2016 Nanostructured disaggregated powders of hydroxide and aluminum oxides *J. New Refractories* **7** pp 37-42
[7] Mikhailov V I 2016 Obtaining and physicochemical properties of materials based on nanodispersed aluminum and iron oxides *Diss. Cand. Chem. Sciences* p 129
[8] Study of phase formation in Fe2O3-H2O and Fe2O3-Al2O3-H2O systems Access mode: http://www.allbest.ru
[9] Toropov N A, Borzakovsky V P, Lapin V V and Kurtseva N N 1965 *State diagrams of silicate systems. Release one. Binary systems: reference book* (Moscow - Leningrad: Nauka) p 546
[10] Zuev V V 2005 *Constitution, properties of minerals and the structure of the earth (energy aspects)* (St. Petersburg: Nauka) p 400
[11] Dunaev A V and Filippova E M 2017 *Unconventional tribotechnology to increase the resource of automotive equipment. The results of 25 years of development. Monograph* (Moscow: GOSNITI) p 252

IOP Conf. Series: Materials Science and Engineering **996** (2020) 012013 doi:10.1088/1757-899X/996/1/012013

Engineering Guidelines for the recovery of fast wearing parts of process equipment of beet sugar plant by plasma spraying with fusing

S V Kartsev, I N Kravshenko, I S Kartsev, I S Splavskiy and S S Amelin

Mechanical Engineering Research Institute of the Russian Academy of Sciences, Moscow, Russia(IMASH RAN)

kazo61@mail.ru

Abstract. The paper considers the possibility of the recovery of fast wearing parts of process equipment of beet sugar plants and provides a brief description of the recovering technology of these parts by plasma spraying with coating fusing. Technological parameters of plasma spraying with fusing of wear-resistant coatings are considered. On the base of the research work the Engineering Guidelines of recovery of fast wearing parts of process equipment of beet sugar plant by plasma spraying with fusing is provided.

1. Introduction

The real strategy of fail-safe control of process equipment of beet sugar plant is the recovery of fast wearing parts. The recovery of these parts is stated as the one of key propriety areas of cost-effective use of resources. For the wide range of parts the cost price of their recovering is equal to 30...70% of the new one. But the service life of the recovering part is higher due to hardening plasma technology usage [1]. The cost reduction is achieved because the recovering part is considered as the blank with necessary shape with some wear on the working surface. The application of the advanced plasma technologies makes the recovering parts quite similar with the new parts. This fact blurs the line between primary and secondary resources and turns them into alternative ones.

2. Materials and methods

The process of the recovery of equipment part is a complex constructive and technological task which characteristics could be changed during its fulfillment. While designing technological process the analysis of technical, economical, ecological and resource saving aspects should be considered. The surface of the part of the equipment of the beet sugar plants is covered with fat films and other impurities. So recovering parts should be thoroughly cleaned. Usually the synthetic detergents and solvents are used for cleaning. After cleaning equipment part should be inspected. The parts with unacceptable defects or wear should be recovered for further using. The material for plasma spraying should be dried and screened before use. Drying should be performed with periodical mixing at the temperature from 100 to 200 °C over the period of 1 or 2 hours immediately prior to powder injector filling. The powder thickness on the sheet pan should not be more than 20 mm thick. Maximum period of time between the drying and spraying is 60 min. After the drying the powder should be put through the sieve type 016 or type 004 made from the square meshed woven wire cloth in accordance with

GOST 6613-86. Remainder left the on the sieve type 004 should be used for plasma spraying. For coating material selection the special software is recommended [2]. Data dialogue system "Barion 1.0" gives the possibility for optimal choosing of wear resistant material. Prepared powder can be stored outdoors at the temperature plus 25°C and relative humidity 70% up to eight hours. If it is necessary to store powder for a longer period of time it should be dried before using in accordance with the above mentioned process.

3. Results

Coating adhesion to the base material and its operational characteristics depend heavily on cleanliness, configuration and roughness of the surface. Mechanical treatment should be performed to make the necessary geometric shape of the detail and to remove wear patterns. Abrasive blasting should be performed to provide the necessary roughness of the detail surface. Such treatment allows to remove the oxide film from the surface. It activates the surface and provides the chemical bonding between the surface and coating particles. It is recommended to use the aluminium oxide grades 13A or 14A with grain size from 30 to 100 microns (GOST 3647-71) or white cast iron shot grades D4K1 or D4K1,5 with grain size from 0.8 to 1.5 microns (GOST 11964-66) as the abrasives for all types of details. Abrasive blasting should be performed in accordance with next working conditions:

- air pressure: from 0.5 to 0.6 MPa;
- abrasive jet impact angle: from 75° to 80°;
- treatment period of one area: not less than 40 seconds;
- distance from the surface to the nozzle: from 35 to 40 mm;
- compressed air flowrate: from 3 to 4 $m^3 \cdot min^{-1}$.

After the abrasive blasting the surface should be blown with compressed air to remove small abrasive particles. Compressed air should be free of oil and moisture in accordance with GOST 9.010-73. The prepared surface should be matte and have no shiny areas. It is recommended to use the thermo-abrasive device for the surface preparation before the plasma spray coating application. Nowadays the thermo-abrasive treatment of the metal surface is considered to be the most effective. Formed at the combustion chamber the gas-flame torch heats the abrasive particles to 200…250°C. So moisture and organic compounds are removed from the compressed air and abrasive material. Removal of different surface contaminations is performed under the high temperature and with high energy impact on the surface layer and low oxygen environment [3].

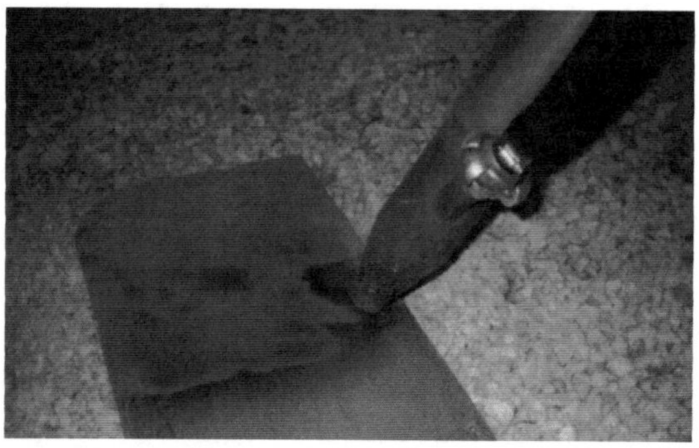

Figure 1. Large surface abrasive treatment with the thermo-abrasive device before the plasma spraying.

The period between surface treatment and spraying should be as minimal as possible and should not exceed two hours because the surface is active and easily oxidizes. During the storage after abrasive treatment attention should be paid to avoid surface contamination. It is not permitted to touch the detail by hands and recommended to use degreased tools and instruments. Uncovered areas of the

IOP Conf. Series: Materials Science and Engineering **996** (2020) 012013 doi:10.1088/1757-899X/996/1/012013

detail located near the spaying zone should be protected by metal screens. The screens should be installed at the distance a bit greater than thickness of coating surface.

Coating should be applied at the special production area with the ambient temperature not less than plus 10°C and humidity not more than 70%.

It is reasonable to use the portable multifunctional plasma unit [4]. Condition of the plasma torch electrode (nozzle, electrode) should be controlled before using. Local wear of jet surface could be not more than 2 mm. Set-up procedure of the portable multifunctional unit should be provided in the following sequence:

- pour prepared powder into the injector dosing device;
- set values for the pressure of gas and cooling liquid and their starting flowrate in accordance with the manual;
- turn on the unit with switched off power supply source, check the correct supply of plasma gas and the oscillator.

Start-up of the portable multifunctional unit is performed in accordance with the procedure shown at the cyclogram (figure 2). Composite powder is spraying.

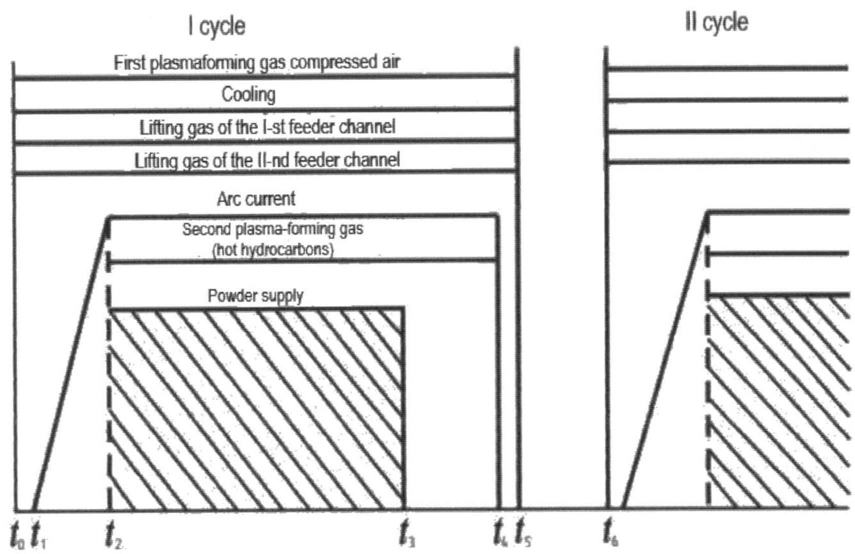

Figure 2. Cyclogram of operation mode of a plasma multifunctional plasma unit in a mixture of compressed air with hot hydrocarbons from the internal combustion engine: t_0 – start of the spraying process, switching on the gas supply system and the free-standing cooling unit;$t_0 + t_1$ – pause for 2...10 seconds (adjustment of gas and coolant flowrates); $t_1 + t_2$ – starting of the operation of the power supply; t_2 – switching on the second plasma-forming gas and the powder feeding; $t_2 + t_3$–time of coating; t_3 – turning off the powder feeding; $t_3 + t_4$ – pause for 2...5 seconds (cleaning the service lines from powders); t_4 – switching off the plasma arc and the supply of the second plasma-forming gas; $t_4 + t_5$ – pause for 2...5 seconds; t_5 – switching off the gas supply system and the free-standing cooling unit; $t_5 + t_6$ – period of part replacement.

It is necessary to turn up the plasma torch while setting up the equipment so that the angle between the jet nozzle and the detail surface would be from 60 to 90 degrees. As the unit switched on the spraying process data should be installed.

During the coating process:

- arc current rate: 170 A;

- plasma-forming gas flowrate: 2.7 m^3 per hour, under the pressure:0.2 MPa;
- spraying distance: 150 mm.;
- plasma torch speed: 40 cm per minute.

Plasma torch speed relative to the coating surface should be determined in a way that one pass coating thickness should be not more than 200 microns. It is recommended to apply an intermediate layer with the thickness not more than 0.05...0.15 mm in one pass to achieve the hard adhesion between the coating and the detail surface. The necessity of the intermediate layer application should be determined separately for the specific case and depending on the detail recovery specification [5].

Depending on the ratio of the compressed air and hot hydrocarbons from the combustion engine in the plasma-forming mixture plasma spraying of powdered materials is performed in reducing, oxidizing or neutral environment that is determined by the α value calculated by formula (1):

$$\alpha = 0.0827 \cdot \frac{Q_v}{Q_{gu}} \tag{1}$$

where:

Q_v – quantity of the compressed air supplied to the mixing unit before the plasma torch, m^3per hour;

Q_{gu}–quantity of hot hydrocarbons from the combustion engine supplied to the mixing unit before the plasma torch, m^3 per hour;

if $\alpha=1$ (stoichiometric composition), neutral environment;

if $\alpha>1$ oxidizing environment;

if $\alpha<1$ reducing environment.

Main coating should be applied layer by layer up to required thickness (figure 3). The thickness of layer to be applied in one pass should not exceed 0.2 mm. Overall temperature of detail during the spraying should not exceed 200°C. In case of specific oxidation tints appeared on the surface spraying should be stopped and detail should be cooled. To prevent the detail overheating the blowing of detail during the spraying with the compressed air in accordance with GOST 17433-80 is allowed. In case of the drop-shaped laps, accidental point-like inclusions (from the plasma torch) forming on the coating the spaying should be stopped. The formed defects with the diameter more than 2 mm should be removed with the sharp knife or scraper.

Figure 3. Plasma spraying of wear-resistant coating.

Depending on the surface shape and the detail size the coating application may be performed mechanically or manually. During plasma spraying first one or two passes are performed without the powder injection in order to heat the detail surface up to 70... 200°C. If the surface is preliminary blasted with the thermo-abrasive device this operation is skipped.

To improve adhesion between the coating and the base material and to reduce residual stresses in coatings appeared as a result of the detail heating with plasma jet, the coating fusing with plasma jet without the spraying material supply is used. Applied coating fusing is performed with the same plasma torch (figure 4). After the unit starting at the beginning of the process the fusing modes should be set:

- arc current of plasma torch: 200 A;
- plasma-forming gas flowrate: 2.7 m^3 per hour at the pressure of 0.2 MPa;
- fusing distance: 40... 50 mm;
- plasma torch speed: 15... 18 cm per minute.

Figure 4. Plasma fusing of wear-resistant coating.

The detail is preliminary heated to the temperature of 300…500°C with plasma jet, and then the plasma torch is stopped on the small area of the coating, which is heated to fusing. The moment of fusing to be detected by shiny mirrored surface appearance, which reflects the plasma jet, after that the plasma torch should be quickly moved to the next area. The coating quality to be controlled visually, there should not be any chips, blisters, peeling, piling and oxidation tints. The presence of cracks should be determined with the ten-power magnifying glass. In case of unsatisfactory coating determination it should be removed by mechanical treatment. The recoating should be performed in accordance with the recommendations mentioned above.

4. Conclusion

As a result of the performed researches the sustainable modes of the plasma spraying with the fusing with the powdered hard alloys were reasoned and determined, which allows to use it for the development of specific technological processes of recovery of worn details of process equipment of the beat sugar plant.

References

[1] Glinsky M A 2018 Development of a method for applying plasma coatings to restore parts of processing equipment of the agro-industrial complex: *dis. of candidate of technical sciences: 05.20.03 Moscow* p 239

[2] Dialog Information system Baryon 1.0, 2008 Selection of materials for coating by gas-thermal methods [Electronic resource] Applied program (624 KB) *Moscow* CD-ROM

[3] Kartsev S V, Petrov A N and Gavrilov D A 2008 Method of surface preparation for plasma deposition of coatings. The patent for useful model RU № 2329886 – *application No.* 2006140518/02. *Bulletin No.* 21.

[4] Kartsev S V, Kravchenko I N and Gurtsiev A Yu 2007 Unit for plasma spraying. Utility model patent RU no. 66341. – *Application no.* 2007113492/22, *Bulletin No.* 25. – **7** C

[5] Kravchenko I N, GlinskyM A, Kartsev S V, Korneev V M and Abdumuminova D T 2020 *Resource - saving plasma technologies for the repair of processing equipment* (Moscow: INFRA-M) p 200

The mechanism of fretting corrosion of details from alloy with shape memory effect

D U Khasyanova

Mechanical Engineering Research Institute of the Russian Academy of Sciences, Moscow, Russian (IMASH RAN)

dinara.khasyanova@mail.ru

Abstract. The article presents the fretting wear features of details from alloys with shape memory effect as applied to pipe connections using couplings from alloy with SME (shape memory effect); describes the mechanism of fretting corrosion in the thermomechanical connections using couplings alloy based TiNi with shape memory effect and superplasticity (SU).

1. Introduction

The fretting wear or fretting corrosion is a common type of wear that accompanies the operation of many contact connections that are subject to small tangential displacements during operation. For the appearance of this type of wear sufficient relative movements of surfaces with an amplitude of 0.025 microns are sufficient. Fretting wear occurs in rivet, threaded, spline, keyway and pin joints, fit parts with tension, steel ropes, hinges, couplings, springs, valves, electrical contact regulators, Cam mechanisms, automatic helicopter screw skewers, gas turbine engine details. The greatest interest is represented by pipeline connections using couplings from alloy with shape memory effect (nitinol), since during experimental testing of pipeline connections using TMC (thermos-mechanical connection) couplings, it was found that the using of damping elements in structures can significantly smooth the transition zone of stress and reduce the processes of fretting corrosion. However, these experiments are staged and haven't deep development.

2. Features of occurrence of fretting wear

The fretting wear characterizes by difficult wear particles removal from the contact zone and a large role of chemical processes in the formation of wear products (fretting corrosion) and activates by cyclic mechanical influences. Fretting corrosion is a phenomenon occurs during the movement of contact surfaces relative to each other and corrosion-mechanical wear of the pressed surfaces during slightly oscillating movements. Fretting wear is a special form of corrosion in terms of the conditions of occurrence and the nature of the manifestation, which differ from the usual friction that occurs during unidirectional sliding. The differences are the speed of relative movement of the contact surfaces in comparison with sliding friction is significantly less and does not exceed 5 $mm \cdot s^{-1}$.

The amplitude of movement of the contact surfaces during the occurrence of fretting is within 0.03 mm, which significantly complicates the removal of wear products from the contact centre, and the presence of oxygen in contact with a fresh and newly developed active surface forms oxides that aggravate damage.

Most often, fretting occurs when press fitments, keyway, rivet, screw, bolt and similar connections, including couplings, which include pipeline connections with TMC. Schematically, the process of creating connections with TMC is shown in figure 1. The connected pipelines must be in an elastic-plastic condition [1].

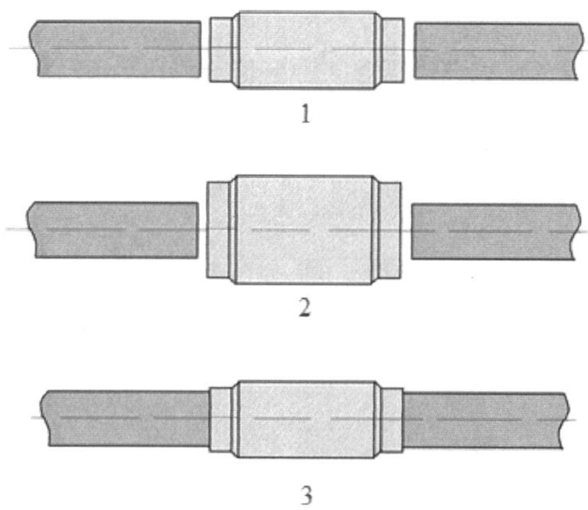

Figure 1. Scheme of formation of all-in-one connections by TMC couplings: 1 – initial condition: the inner diameter of the coupling is less than the outer diameter of pipes; 2 – deformation coupling diameter: inner diameter of the coupling larger than the diameter of the pipes; 3 – finished joint: after Assembly of the clutch restores its shape, is reduced in diameter and tightly compresses the pipe ends.

The using of all-in-one thermos-mechanical connections (TMC) with couplings from alloys with shape memory effect (SME) has the following advantages:

- the strength of the connections is determined by the strength of the pipe in the supply, including heat-strengthened;
- the endurance limit of the connections is within 0.9-0.96 of the endurance limits of the connected pipes;
- the complexity of installation and assembly work is less than 2 minutes;
- high density of installation, including in hard-to-reach places;
- corrosion resistance of TMC couplings is not lower than stainless steel and titanium alloys;
- used for connecting pipes of various thicknesses, including thin-walled (from 0.2 mm), allowed a combination of different brands of materials.

The intensity of fretting formation depends on the amplitude of relative displacements determined in the range from 8×10^{-7} mm. For steel, the most intense fretting occurs at amplitudes of 0.1-0.2 mm [2]. The intensity of wear depends on the amplitude and the types and contacted materials properties, stresses, surface micro relief, and operating conditions.

Fretting phenomena of intergranular corrosion can occur even at low stresses when combined with minuscule amplitudes for such materials that are prone to sticking and picking, as well as those circumstances that almost all metals and refractories are dissolved in titanium and its alloys.

One of the features of the surface properties of titanium alloys, which significantly distinguishes them from alloys based on many metals, is the high reactivity of the surface, which causes a low load of contact setting during friction, high values of the friction coefficient and wear rate. The sliding friction of titanium on any metal leads to sticking to its surface and the material seems to get stuck caught in a sticky layer of titanium, which leads to rapid wear.

Factors contributing to the manifestation of fretting: establishment of a time factor for specifically contacted materials that contributes to the violation of the continuity of the surface layer, leading to deep breakouts of the material and depending on the amplitude and frequency of relative displacement, chemical activity of surface energy, fatigue strength determined by the number of cycles and operating stresses.

Factors contributing to the reduction of fretting: design solutions for creating such stress Conditions that reduce or increase the amplitude of mutual movements from the zone of 0.1-0.2 mm, the using of damping elements that prevent vibration in the contact zones.

Titanium nickelide is consist of half of titanium like all titanium alloys has a tendency to stick and is character by a high coefficient of friction in both the austenitic and martensitic conditions. In addition, titanium nickelide exhibits SME and SU, and has a high ability to move each other in the contact zone, up to 8% [3].

Taking into account the fact that in this issue the goal is to fully consider the fretting corrosion of TiNi alloy, it should be noted that the causes of fretting corrosion, such as accelerated metal oxidation due to continuous removal of the oxide layer and metal wear due to friction are the most important [4].

Titanium nitride is a golden yellow film that has a high microhardness, but does not show sufficient plasticity, which does not allow maintaining the integrity of traditional protective coatings. Depending on the production method, the ratio of nitrogen atoms per titanium atom can vary in titanium nitride from 0.6 to 1 [5].

3. The mechanism of fretting corrosion

The main driving force in the process of fretting corrosion is friction [6]. The interaction of surfaces during friction is concentrated in a certain volume of material endowed with elastic-viscous properties. This volume can be considered as the so-called "third body". When the surfaces move relative to each other, the surface layer is destroyed and a "third body" is formed at the same time. As a rule, the contact zone of two surfaces is usually characterized by the presence of an intermediate layer between the solids, which is a film (lubricant) of oxide, adsorbed water steam and a film of degraded material. The area consisting of the film and the degraded coating material is called the "third body". A generalized approach to the mechanism of corrosion of titanium alloys allows using to distinguish three main stages of development of fretting corrosion of metal and its alloys under friction conditions. At the first stage of the fretting process, there is a hardening of the contacting surfaces of the material and cyclic fluidity of the thinnest surface layers, most of the protrusions interact with each other plastically. This is facilitated by the setting of metal in contiguous micro-surfaces after the destruction of natural oxide films. Protrusions that have collapsed due to fatigue and cut-off set points create primary products of destruction, some of which are oxidized. However, most of the wear products at this stage of fretting are metal particles. The transition of the surface layers to the dispersed and ultra-disperse conditionals accelerates the oxidation reactions. At the second stage of fretting corrosion, fatigue damage continues to accumulate in the surface layers. Simultaneously with fatigue damage in the form of an ultrafine layer of metal particles and their oxides, a corrosive environment is formed in the friction zone due to adsorption on oxygen oxides. The rate of wear at this stage is low, and wear is mainly associated with the destruction of oxide films formed on the friction surfaces in those areas where there is an intensive destruction of surface oxide films, setting processes are intensively developed. The intensity of their development is further increased due to the fact that during the formation of wear products, the working area of the contact surfaces is sealed. Accordingly, the access of oxygen to this zone is practically stopped, and therefore the processes of formation of oxide films are sharply slowed down, and contact is made on clean surfaces that are more prone to setting. At this stage, there is a special (mechanical and chemical) mechanism for intensifying the oxidation of metal surfaces, due to the fact that during alternating contact interactions in the thinnest surface layers, a fine structure appears that is prone to active oxidation. Later, in the surface layer, a mixed structure is formed - from the coating metal and oxides of this metal, which can play a protective role, reducing the wear rate. In the second period of fretting corrosion, under optimal conditions of the formed contact, the pre-hardened coating layers experience more moderate cyclic loads, but further accumulation of fatigue damage occurs in them, which is aggravated by corrosion processes. High-dispersion oxides formed during fretting corrosion of metals, being semiconductors, give the process a catalytic character [7]. This effect of oxides is manifested in accelerating the processes of oxygen adsorption in active radicals and ion-radical forms. In this case, a reaction electrolytic medium is formed between the contacting surfaces of the electrical contacts. The third stage of fretting corrosion is associated with the final destruction of the damage zones of contact surfaces that have already been loosened by fatigue and corrosion processes. Taking into account the possibility of electrochemical processes, this stage can be called the stage of corrosion-fatigue failure. During this period, the surface layers of metal that have been subjected to cyclic deformations for a long time become so softened that they lose their stability, and their progressive separation begins, which is manifested in the rate of wear. The mechanism of destruction of the surface layers of the contacting surfaces is shown in figure 2.

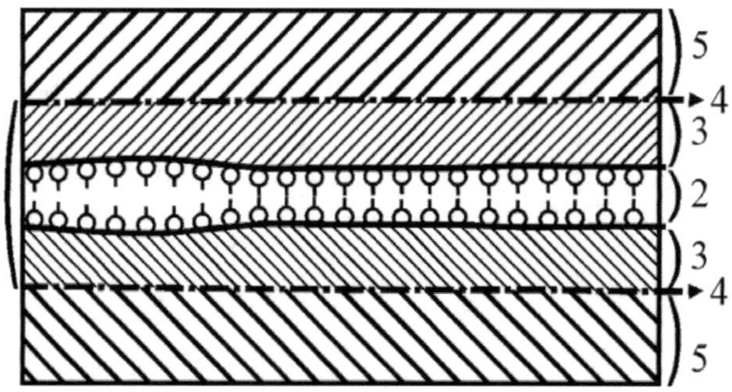

Figure 2. Contact diagram of two bodies: 1 — "third body"; 2 — adsorbed layer; 3 — oxides and other chemical compounds; 4 — loosened layer; 5 — the main coating material.

During experimental testing of pipeline connections with TMC couplings, it was found that the use of damping elements in structures can significantly smooth the transition zone of stress and reduce the processes of fretting corrosion.

However, these experiments were considered as staged and did not have deep development. On the surface of samples from titanium alloys damaged by fretting, the consequence of mechanical action on the surface of a solid abrasive and the impact of a viscous amorphous medium formed by the study of the physical and chemical properties of damage products formed during fretting on the contact surfaces cannot be considered objective, since they are rather a consequence of the process, and not a real factor in fretting.

More reliable testing of both structural elements of TiNi damping devices and materials that show and do not show SU during operation, as well as modes and methods of applying titanium nitride (sputtering or gas nitriding) is necessary. Consider the influence of these processes on the fatigue strength of materials, as well as the use of surface films made of colloidal graphite to prevent oxidation of transition zones between titanium nitride and TiNi [8].

4. Results

The mechanism of corrosion in TiNi-based alloys with shape memory effect is described, which identifies three main stages of development of fretting corrosion of metal and its alloys under friction conditions. In the course of the work, it was found that on the surface of samples from titanium alloys damaged by fretting, the consequence of mechanical action on the surface of a solid abrasive and the effect of a viscous amorphous medium formed by the study of the physical and chemical properties of damage products formed during fretting on the contact surfaces cannot be considered objective, since they are a consequence of the process rather than a real factor in fretting.

5. Discussion

Nitinol has a high complex of mechanical properties. The high corrosion resistance of titanium and its alloys is due to the formation of a protective oxide film on their surface, which prevents the release of ions into the medium. However, titanium alloys have low wear resistance, which makes them vulnerable to fretting corrosion. The mechanism of fretting corrosion in TiNi-based alloy with shape memory effect compounds presented in this paper gives an idea of the sequence of fretting corrosion development, but does not take into account all the factors that affect this process. Therefore, in future work, it is necessary to study the factors that affect the intensity of fretting corrosion, such as temperature, composition and corrosion activity of the medium, the value of the specific load, and the number of loading cycles. To analyse what methods can be applied to improve tribological characteristics and resistance to fretting corrosion of titanium alloys.

6. Conclusion

The paper presents the features of fretting wear of parts made of an alloy with the shape memory effect in relation to pipeline connections using couplings made of an alloy with SME. The mechanism of training-corrosion in TiNi-based alloy compounds with shape memory effect is described.

References

[1] Khasyanova D U 2019 *J. of Mach. Manufacture and Reliability* vol 48 issue 1 pp. 88-92
[2] Waterhouse R B 1976 *Fretting corrosion* (Moscow: Mechanical engineering) p 272
[3] Khmelevskaya I, Prokoshkin S, Inaekyan K, Gurtovaya I, Korotitskiy A, Brailovski V, Demers V and Dobatkin S 2008 *Functional properties of Ti Ni based shape memory alloys* CIMTEC (Sicily) pp 156-61
[4] Clayton P 1993 Tribological behavior of a titanium-nickel alloy vol 162-164 issue part A pp 202-10
[5] Tomashov N D and Altovsky R M 1963 *Corrosion and protection of titanium* (Moscow: Mash-Giz) p 277
[6] Batrakov V P 1957 *Questions of the theory of corrosion and passivity of metals in oxidizers* (Moscow) p 591
[7] Semenova I V, Florianovich G M and Khoroshilov A V 2006 *Corrosion and protection from corrosion* (Moscow: FIZMATLIT) p 376
[8] Zhuk N P 1976 *Course of the theory of corrosion and protection of metals* (Moscow: metallurgy) p 472

IOP Conf. Series: Materials Science and Engineering **996** (2020) 012015 doi:10.1088/1757-899X/996/1/012015

Modelingof stress-strain state of a pneumatic tire in Abaqus software at stationary rolling

P V Korolev[1,2], M A Shilov[1,2], L B Maslov[1,2] and S V Fomin[1]

[1] Vyatka State University
[2] Ivanovo State Power Engineering University

vedunking@mail.ru

Abstract. The paper presents a step-by-step development of a finite element model of stationary rolling of a radial automobile tire on a surface, where friction is taken into account. The method of finite element modeling (FEM) is used to determine the influence of the method of setting the rolling of a pneumatic tire on its performance properties. The authors come to the conclusions about the advisability of using various methods for creating models of pneumatic tires and give recommendations for optimization of the numerical experiment.

1. Introduction

Currently, physical and mechanical properties of pneumatic tires by experimental methods are studied well enough and are actively used in predicting the performance characteristics of materials [1]. The growth of the car park and its modernization lead to increasing average speeds, traffic intensity and the growth rates of road accidents (RTA). In this regard, the problem of ensuring traffic safety on the road is urgent. Car manufacturers are constantly trying to improve their design in order to improve safety performance. However, they may not be effective enough if there is no reliable connection between the tire and the road due to their grip properties. An increase in vehicle speed leads to increased tire tread wear. When the tread is worn, the adhesion of the tire to the road surface decreases sharply, the braking distance of the car increases, the parameters of stability and controllability deteriorate. Therefore, the purpose of the research is to predict the performance properties of tires by numerical methods. The linear model of the dependence of stress on deformation of some materials can not be described under mechanical loading. Rubber belongs to this group [2]. It can have large deformations, that is why it is referred to the category of hyperelastic materials [3] and requires choosing a special model when performing engineering calculations.

Rubber is nonlinear; therefore, there are special models which can describe such hyperelastic behavior of materials [4]. Choosing a model is the main issue when performing calculations.

Figure 1 shows the deformation range for each mathematical model.

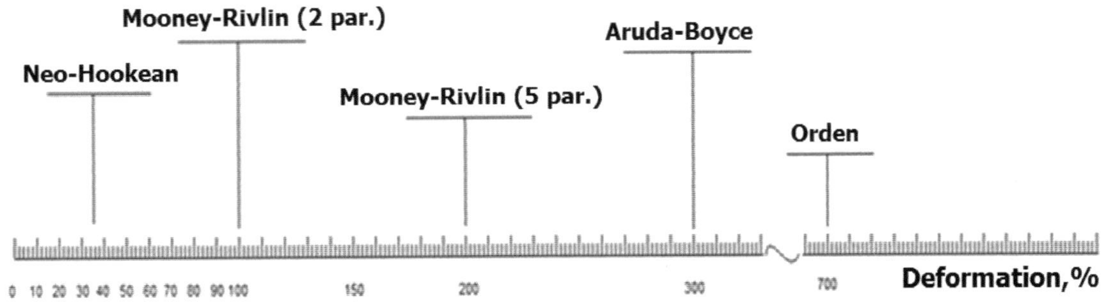

Figure 1. Application range of models of hyperelastic materials.

The Neo-Hookean model is the simplest one. It is one of the first to describe the behavior of the stress-strain state of rubber [5]. However, it is suitable only for small deformations (up to 30–40%) [6]. The two-component Mooney–Rivlin model is also widely used for deformations up to 100%, although it is not suitable for describing the influence of stiffness effects [7]. For deformations within 200%, it is better to use the five- and nine-component Mooney–Rivlin models [8]. The polynomial model also better describes cases of large deformations. Yo three-component model suits for describing the behavior of rubber at high elongations, but is not suitable for small deformations [9]. Moreover, the required material parameters are rather difficult to obtain. The Arruda-Boyce and Gent models are suitable for cases of both small and large elongations, but it is best to use them for deformations within 300% [10]. The Ogden's model is based on principal elongations and gives a very good curve fit, but requires a lot of computational resources and is used for deformations more than 700% [11]. The Blatz - Ko model is used for compressible foamed polyurethane rubbers [12]. Other highly compressible elastomeric foams are described by the hyperpenic model. The two-parameter Mooney-Rivlin model is best suited for solving the problem [13]:

$$W = c_{10}(\bar{I}_1 - 3) + c_{01}(\bar{I}_2 - 3) + \frac{1}{d}(J_{el} - 1)^2 \quad (1)$$

where the coefficient of incompressibility of the material d (Pa^{-1}) is calculated taking into account Poisson's ratio (ν):

$$d = \frac{1 - 2\nu}{c_{10} + c_{01}} \quad (2)$$

where c_{10} and c_{01} (Pa) are the model coefficients.

Materials with highly elastic deformation (rubber resin, rubber, some plastics, and textiles, which are capable of significantly greater stretching under uniaxial loading than, for example, steel and various metals) show the linear dependence σ—ε only at very small initial sprains. In general, despite the high reversibility of deformation, the dependence σ—ε for these materials is not linear and is usually not monotonic. Consequently, materials that do not correspond to the well-known Hookean position cannot be characterized by a single constant value of the modulus of longitudinal elasticity E calculated from the conditional stress σ. In the section of nonlinear dependence, the modulus of the material E can be determined only in differential form. The local module, which is sometimes used, does not provide a structurally meaningful assessment of the material. The evaluation of rubber properties with a conventional stress, which is used in laboratory practice, corresponding to tension by 100, 300, or 500% against the initial length of the sample is also inconsistent. These so-called "modules" are only the ordinates of some intermediate points of the curve σ—ε, but not material constants [14-19].

2. Materials and methods

The object of research was a radial automobile tire. This type of tire has the carcass threads arranged radially along the tire profile from one bead to the other. They are parallel to each other in all layers. Only the breaker has a diagonal design. The radial arrangement of the carcass threads does not allow the rubber to stretch strongly in the transverse direction, and the breaker keeps from the longitudinal movement. Such an arrangement of the carcass threads reduces pressure approximately by half. Therefore, it becomes possible to reduce the number of layers of cord, due to which the weight of radial tires is less than that of bias tires [20]. The carcass of radial tires is more elastic due to the lower thickness and has less internal friction. Consequently, a working tire of this design generates less heat [21]. It allows to increase tread thickness, tread depth and tire life. The breaker, on the other hand, is very rigid and practically inextensible in the radial direction. Radial tires also have greater stability of the shape of the contact patch with the road surface, create less rolling resistance, and thus provide lower fuel consumption [22].

At present, stress-strain state is described both by analytical and numerical methods. To obtain numerical data, the following software systems (PC) are used: Ansys, Abaqus, Nastran, etc. We have chosen the finite element modeling PC Simulia Abaqus from Dassault Systemes as a modeling tool due to:

1) wide opportunities for nonlinear analysis;

2) the ability to use an independent library of materials and elements.

There are many classifications of car tires. We use the following devision:

- by sealing: chamber and non-chamber;
- by design: radial and diagonal;
- by size: small, medium, large;
- by type of profile: toroidal, wide-profile, low-profile, arched, pneumatic rollers.

In this classification, the most massive class is radial medium-sized pneumatic tires of toroidal profile, which is the object of study in [23]. The subject of the research is numerical methods for modeling complex mechanical systems.

Tests of rubber samples were carried out in accordance with GOST 270-75 "Method for determining elastic-strength properties in tension". The standard applies to rubber and establishes a method for determining elastic strength properties in tension in terms of: tensile strength, elongation at break, stress at a given elongation. The essence of the method is to stretch the samples at a constant rate to rupture and measure the force at given elongations, at the moment of rupture and elongation of the sample at the moment of rupture [24].

3. Results

Finite element modeling and visualization software packages are a convenient tool for the numerical study of tire characteristics and tire wear. In our research, a finite element model of a car tire was obtained by means of the Abaqus program complex for finite element strength calculations [25]. We used a pneumatic tire as a prototype, its structure is shown in figure 2.

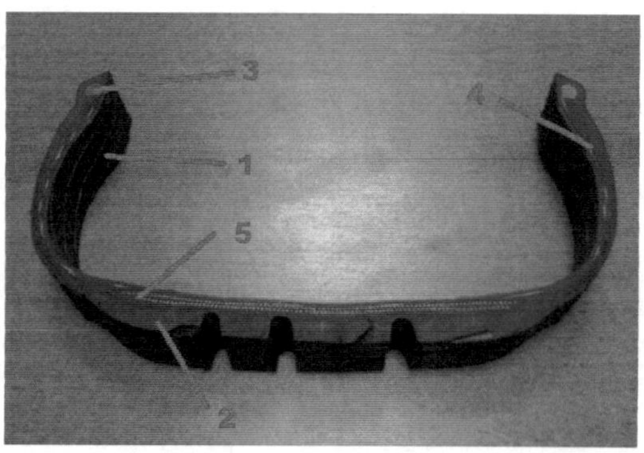

IOP Conf. Series: Materials Science and Engineering **996** (2020) 012015 doi:10.1088/1757-899X/996/1/012015

Figure 2. Tire structure: 1) hermetic layer; 2) protector; 3) steel bead; 4) carcass; 5) breaker.

The geometric model of a radial car tire consists of five distinct parts: a rubber tread, metal rings, a nylon two-layer bead, a two-layer breaker, and a hermetic layer (figure 3).

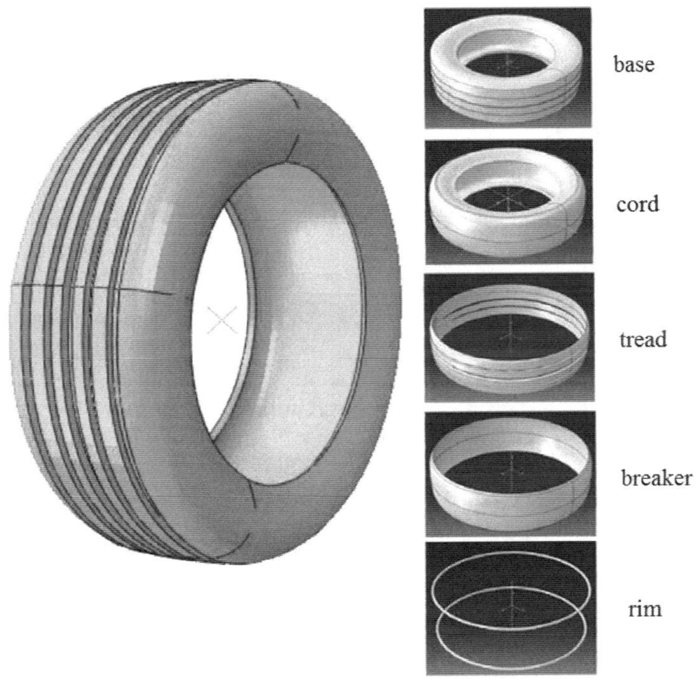

Figure 3. The model and its components.

The nylon threads of the carcass bead are located perpendicular to each other in two composite reinforcing layers of the tire: along and across the axis of rotation of the tire. Similarly, the metal strands of the breaker, which are a composite in a rubber filler, are laid diagonally at an angle of 65 and -65 degrees to the axis of rotation of the tire.

SKI-3 rubber, the properties of which we obtained earlier [26], was taken as the material of the rubber filler of the tire. Non-linear properties of nylon thread are taken from literature sources. The properties of nylon threads and rubber are specified by a hyperelastic material model. The materials of the rim rings and metal threads of the breaker layers are taken to be linear. Determining the properties of the supporting beads is simplified by introducing the assumption that the geometry of the support is considered independent of the material surrounding it. Thus, the finite element mesh for the beads was constructed independently of the layers of the rubber components of the tire. For this, Abaqus provides surface elements, which help to describe the properties of support. The surface features had no structural properties and were only used to define the geometry of the beads. Such elements are embedded in 3D solid elements used to model the rubber components of a tire. At the same time, there are restrictions between the nodes of the mesh of the matrix of the rubber material and the nodes of the surface elements of the support. The use of embedded elements prevented the difficulties of mesh construction for support, such as very small elements between the layers. To select a mathematical model of a material, for example, rubber or nylon, we calculated the selected coefficients according to the results of experimental tensile data [27] .

A 3-dimensional, 8-node hybrid finite element C3D8H was selected as the tire finite element (figure 4). M3D4H 4-point element was selected for bead and breaker components.

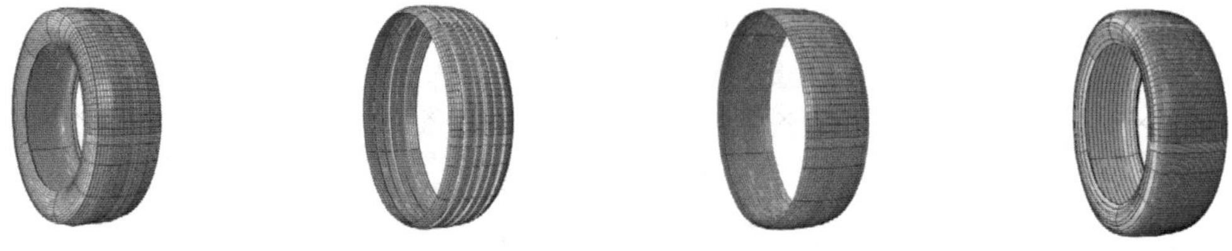

Hermeticlayer Protector Breaker Bead

Figure 4. Breakdown each tire component into finite elements.

In the axisymmetric formulation, the geometry of the filler was divided by elements of the CGAX4 type: a 4-node bilinear axisymmetric element, further transformed with the help of the SYMMETRIC MODEL GENERATION module into a C3D8H element as a result of "turning" the axisymmetric model around the axis of rotation. Similarly, the MGAX1 type elements were transformed: a two-node linear axisymmetric membrane element into an M3D4H type element.

The number of elements for the axisymmetric model was 653, for the 3D model - 47881.

The inline element method was used to specify the inlining of an element or group of elements into "host" elements. An inline element algorithm was used to model the support. Geometric relationships between inline element nodes and host elements were found using Abaqus. If an inline element node was inside a host element, the translational degrees of freedom in the node were eliminated and the node became an inline node. The translational degrees of freedom of the built-in node were limited by the interpolated values of the corresponding degrees of freedom of the main element. Built-in elements could have rotational degrees of freedom. For modeling a car tire, the elements of the bead and breaker were built into the elements of the hermetic layer.

4. Discussion

The numerical experiment consisted of a dynamic rolling of a pneumatic tire model on an analytical plane with a simplified tread pattern. All parts of the model were combined into a prefabricated unit.

Static loading of the tire had three stages:

Stage 1 - tire inflation;

Stage 2 - pressing by shifting the analytical plane to the tire in a coordinate way in order to set the interaction in the contact zone;

Stage 3 - applying a weight load to the tire.

A universal (direct) approach is to build a full 3D tire model ready for all types of testing. It is adapted to rapidly changing boundary and initial conditions, geometry and properties. The disadvantages of this method are the heaviness of the model (15000-20000 elements) and the duration of the calculation (from 15-20 minutes for static problems to 2 hours for dynamic problems).

At the same time, Dassault Systemes proposes a methodology for the numerical test of tire rolling, which has several stages:

- building a tire model in an axisymmetric setting;
- using the Dassault Systemes plugin, which implements the algorithm of actions;
- transformation of the section into a three-dimensional view in an axisymmetric setting, pressing it to the surface;
- obtaining a complete static three-dimensional model of a tire with an axisymmetric stress-strain state;
- obtaining a stress-strain state (SSS) during steady rolling (Steady State Transport).

Steady-state rolling occurs at a constant speed of the wheel with a coefficient of friction between the wheel and the surface of 0.5. It is usually calculated on a simplified model, where the part of the tire (fitted to the wheel together with the rings) is cut off. It is replaced by its binding to the reference point (RP).

Analysis of the experimental results showed a small deviation between the test data of the full 3D model and two variants of axisymmetric models transformed using the plug-in. It allows to use all three methods of constructing the wheel. Taking into account the large amount of time spent on the numerical calculation of the three-dimensional model, with simple shapes of the tire tread, which can be obtained by rotating the sectors around the tire axis, it is more economical to use the axisymmetric model transformed with a help of plug-in. For a complex tread, without the possibility of obtaining sectors by axial rotation of a flat section, it is possible to use only a complete construction of a three-dimensional tire model.

Modeling results are shown in figure 5. The solution to the problem of stationary tire rolling on a flat surface with allowance for friction was carried out in a hybrid Euler-Lagrangian formulation.

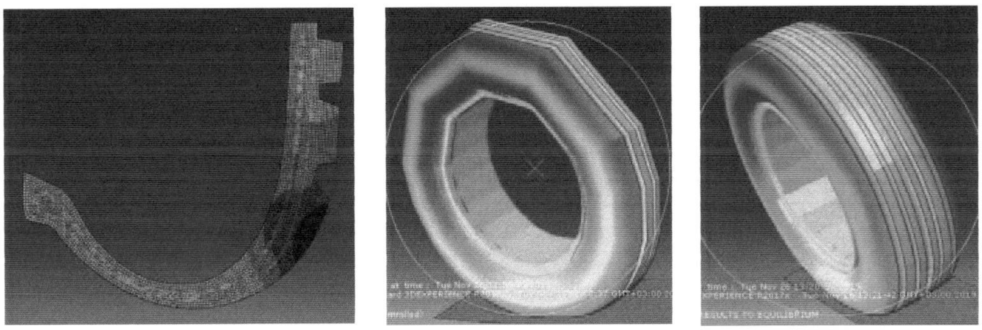

Figure 5. Results of the tire rolling modeling stages.

To study the effect of each type of load on the stress-strain state, we set several loading modes, where the first was a test one, and in each subsequent one we changed only one tire load parameter. Loading data is presented in table 1.

Table 1. Loading modes.

Parameters Dimension	Inflation MPa	Pressure N	Angular velocity rad·s^{-1}	Linear velocity mm·s^{-1}
1	0.3	3300	74	22222
2	0.2 0.4	3300	74	22222
3	0.3	2800 3800	74	22222
4	0.3	3300	50 100	15000 30000

For each loading mode, a numerical experiment was performed in all three stages. The value of the maximum contact pressure in the tire was selected as the measured value. By measuring the value of the parameter for each experiment, we obtained a set of values of the contact pressure of numerical experiments. To assess the degree of influence of each parameter on the magnitude of the contact stress, we took the change in each parameter in percentage terms, constant for each experiment, and evaluated the change in voltage in the contact zone. Taking as a basis the change in the tire pressure value, which was about 15%, the boundaries of the considered sections were obtained for the remaining parameters. The calculation results are shown in table 2. Putting the values of significant areas along the Ox axis, we obtained the range of variation of the contact pressure values for each experiment (figure 6).

Table 2. Results of calculating the boundaries of significant areas.

Contact pressure, MPa	Inflating tires, MPa	Area boundaries, MPa
0.887	0.2	0.254
0.816	0.3	0.3
0.736	0.4	0.345
Contact pressure, MPa	Tire pressure, N	Area boundaries
0.641	2800	1.178
0.816	3300	1
0.95	3800	0.868421
Contact pressure, MPa	Movement speed, rad·s^{-1}	Area boundaries, rad·s^{-1}
0.824	50	62.787
0.816	74	74
0.732	100	85.212

The analysis of the data made it possible to prove the greatest influence of the weight load parameter on the change in contact pressure in the zone of contact between the tire and the road surface. The pressure values with a change in axle load by 15% more than 3 times exceeded the change with a similar change in the parameters of inflation or movement speed, which corresponds to the data presented in [28].

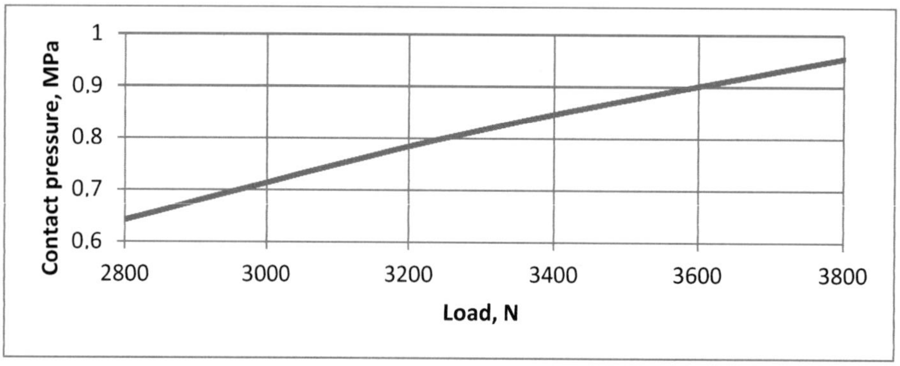

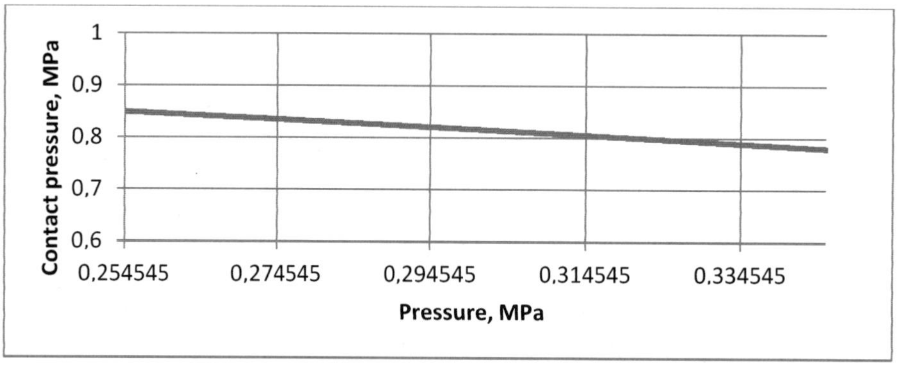

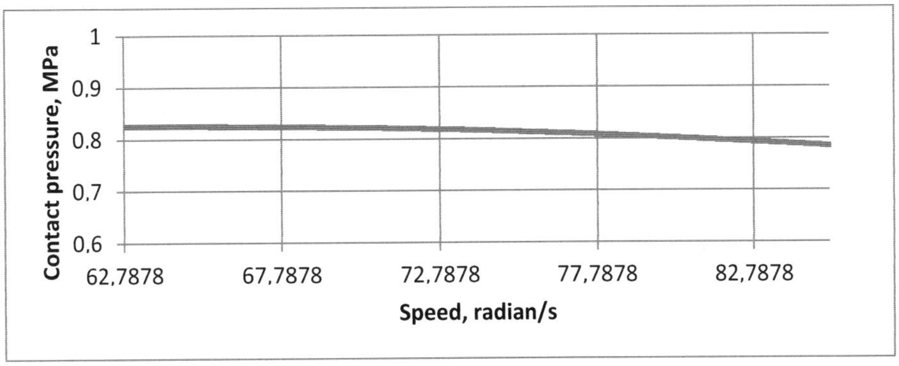

Figure 6. Change in contact pressure when changing input parameters by 15%.

5. Conclusion

During our research, we:

1. Revealed the main structural and physical-mechanical features of the car tire.
2. Investigated the elastic properties of SKI-3 rubber, the results were approximated by the Mooney-Rivlinhyperelastic material model.
3. Developed a finite element model of quasi-static rolling of a car tire on a flat surface.
4. Revealed the greatest influence of the axial pressure parameter on the pressure in the zone of contact of the tire with the surface. A qualitative comparison of the modeling results with other studies showed the correlation of the results.

References

[1] Medveditskov S I and Zadvornov V N 2013 Influence of life endurance and tread wear on the coefficient of resistance to lateral tire drift *Automotive Industry* vol 8 pp 20-22
[2] Zaporozhtsev A V 1972 Influence of car tire wear on basic characteristics vol 2 pp 27-29
[3] Bapat C N and Batra R C 1984 *International journal for numerical methods in engineering* vol 20 pp 1911–27
[4] Oden J T and Lin T L 1986 *Computer methods in applied mechanics and engineering* vol 57 pp 297–367
[5] Oden J T, Lin T L and Bass J M 1988 *Tire science and technology* vol 16 pp 18–43
[6] Le Tallec P and Rahier C 1994 *International journal for numerical methods in engineering* vol 37 pp 1159–86
[7] Nackenhorst U 2004 Computer methods in applied mechanics and engineering vol 193 pp 4299–22
[8] Wriggers P 1995 Archives of computational methods in engineering vol 2 pp 1–49
[9] Guangdi Hu and Wriggers P 2002 *Computer methods in applied mechanics and engineering* vol 191 pp 1333–48
[10] Coulomb C A 1821 Theorie des Machines Simple p 379
[11] Rivlin R S 1997 Some applications of elasticity theory to rubber engineering *Collected Papers of R S Rivlin* vol 1 pp 9–16
[12] Treloar L 1953 Physics of elasticity of rubber *Publishing house of Foreign Lit* p 244
[13] Lurie A I 1980 Nonlinear theory of elasticity *Science* p 260
[14] Kumar N 2016 Hyperelastic Mooney-Rivlin Model: Determination and Physical Interpretation of Material Constants *MIT Int. J. of Mechanical Engineering* vol 6 pp 43-46
[15] Ivanov D V 2008 Determination of constants for Neo-Hookean and Mooney-Rivlin models based on the results of uniaxial tension experiments *Collection of scientific works. Mechanics. Mathematics* pp 114-17
[16] Yeoh O H 1993 Some forms of the strain energy function for rubber *Rubber Chemistry and*

Technology vol 66 pp 754–71

[17] Zhi-Qiang Feng, Magnain B and Cros J 2006 Solution of large deformation impact problems with friction between Blatz *Ko hyperelastic bodies* pp 113–26

[18] Arruda E M and Boyce M C 1993 A three-dimensional model for the large stretch behavior of rubber elastic materials *J. Mech. Phys. Solids* pp 389–412

[19] Qi H J and Boyce M C 2005 Stress-strain behavior of thermoplastic polyurethane *Mechanics of Material v*ol 37 pp 817–39

[20] Gent A N 1996 A new constitutive relation for rubber *Rubber Chem. Tech* vol 69 pp 59–61

[21] Ogden R W 1972 Large deformation isotropic elasticity – on the correlation of theory and experiment for incompressible rubberlike solids, proceedings of the royal society of london *Series A. Mathematical and Physical Sciences* vol 326 pp 565–84

[22] Tretyakov O B Gudkov V A and Tarnovsky V N 1992 Friction and wear of tire p 176

[23] Goryacheva I G, Makhovskaya Yu Yu, Morozov A V and Stepanov F I 2017 Friction of elastomers. Modeling and experiment *Izhevsk Ins-t of comp. researches* p 204

[24] Ryzhov S A, Ilyin K A and Varyukhin A N 2006 Design of tires using the ABAQUS software package *CAD and graphics* pp 77-81

[25] Korolev P V, Shilov M A, Fomin S V and Maslov L B 2019 Finite-element simulation of cyclic compression of a cylinder with account for energy dissipation *Int. Conf. of Young Scientists and Students "Topical Problems of Mechanical Engineering" (ToPME-2019)*

[26] Yufereva Ya N, Maslov L B and Fomin S V 2019 Experimental study of mechanical properties of elastomeric materials based on rubber mixtures *In the collection: XII All-Russian Congress on Fundamental Problems of Theoretical and Applied Mechanics. Collection of works* vol 4 pp 228-30

[27] Korolev P V, Maslov L B, Shilov M A and Fomin S V 2019 *Proceedings XXXI international innovative conference of young scientists and students on the problems of Mechanical Engineering (MIKMUS - 2019)* pp 24-27

[28] Yang X 2011 Finite element analysis and experimental investigation of tyre characteristics for developing strain-based intelligent tyre system *School of Mech. Eng. College of Engineering and Physical Sciences The University of Birmingham* p 220

Comparative analysis of tribological properties of friction bands used in lifting and transport machinery

O I Kulakov

Laboratory of Friction Units for Extreme Conditions, Mechanical Engineering Research Institute of the Russian Academy of Sciences, 4 Bardina str, Moscow, Russia

kulakov@imash.ru

Abstract. The comparative tests of asbestos-containing EM-1 and asbestos-free BAC friction bands were conducted. The results indicate that BAC friction bands provide more stable friction coefficients. The usage of BAC friction bands may reduce maintenance costs for the brakes of powered industrial machines.

1. Introduction

Drum and band brakes are widely used in traveling cranes as they can develop high braking torque and withstand high loads [1]. The stability and efficiency of drum brakes is evaluated by the breaking torque, friction bands temperature and their wear.

Considerable part of friction bands in use is manufactured using the asbestos fiber. The World Health Organization acknowledges asbestos as carcinogenic agent, for instance, the usage of asbestos in the most of developed countries is limited or forbidden [2]. The developing countries are sharing the same path of gradual enhancing of restrictions on use of asbestos in industry and construction.

Considering the possible restrictions on use of asbestos in Russian Federation the need of replacing asbestos-containing bands with asbestos-free ones may arise. Thus, it is reasonable to conduct comparative tests of such materials.

2. Experimental part

A widely used Russian asbestos-containing elastic material EM-1 (GOST 15960-96) and Polish composite friction band BAC were chosen as comparison objects. The BAC band is woven band of ceramic yarns with very thin monofilaments reinforced with brass wire and soaked with composition of synthetic and natural resins. The specifications of these bands are shown in table 1.

Table 1. Band specifications.

	EM-1	BAC
Max. unit pressure, MPa	1.5	1.2
Max. rotational velocity, $m \cdot s^{-1}$	18	20
Max. temperature of constant operation, °C	200	350
Nom. friction coefficient	0.44	0.40

The comparative experimental evaluation of friction and wear behavior of these materials was conducted. The disc-shoe experiment was performed on the modernized II 5018 machine [3] under parameters matched to real-life ones. The photo of II 5018 machine is shown in figure 1.

Figure 1. Modernized II 5018 machine.

The aim of this experiment was to assess the friction coefficient and the wear intensity under different loads.

9 specimens of each material were manufactured. The Ø 50 mm disc counterfaces were lathed from St45 GOST 7417-75. Each specimen was paired to its own disc to get rid of the counterface wear influence. The EM-1 specimens were milled shoe-shaped. The BAC specimens were glued to 1 mm thin St45 shoes using the VK9 compound. The normal load was set to 200, 300 and 400 N. The disc rotational frequency was set to 200, 300, 400 rpm. The duration of each test was 450 sec disregarding the break-in period. The measurement and registration of braking torque and contact temperature was conducted during the tests. The temperature was measured by K-type thermocouples mounted at 45° angles to the contact surface. Additional calculations of friction forces and friction coefficients were made. The weight wear was evaluated using the VLR-200 analytical scales. The linear wear was not evaluated due to inappropriate materials structure. The counterface wear was insufficient and was not taken into account.

The rotational frequencies chosen correspond to low rotational velocities of the friction track (1.046 – 2.095 $m \cdot s^{-1}$). Taking nominal contact area into account, 200 N load approximately corresponds to 1.04 MPa unit pressure, 250 N load – to 1.3 MPa, 300 N load – to 1.57 MPa. It was noted that actual contact area for BAC material is bigger than nominal, so the actual unit pressure was somewhat lower. Thus, at 300 N load the unit pressure was exceeded by 5% and 31% for EM-1 and BAC materials correspondingly. During testing the specimen temperature did not exceed the specifications.

3. Results

The experimental data was analyzed; the graphs of friction coefficients and specimen temperatures were plotted (figures 2-5). The experimental data is shown in table 2.

Table 2. Experimental data.

			M, N·m		T_{final}, °C		$F_{friction}$, N		$F_{friction\ average}$		J, mg·min^{-1}	
	P, N	V, rpm	EM-1	BAC	EM-1	BAC	EM-1	BAC	EM-1	BAC	EM-1	BAC
1	200	200	2.33	2.36	76	89	93.02	94.32	0.46	0.46	0.450	0.500
2	200	300	3.99	1.84	156	74	159.57	73.61	0.75	0.37	1.025	0.369
3	200	400	3.90	1.77	151	59	155.98	70.96	0.80	0.36	1.694	0.359
4	250	200	3.26	2.47	108	78	130.41	98.91	0.52	0.40	1.106	0.231
5	250	300	3.35	2.78	164	82	133.90	111.19	0.54	0.44	2.638	2.981
6	250	400	3.69	2.58	170	99	147.72	103.37	0.60	0.41	2.125	0.625
7	300	200	2.22	2.19	72	69	88.95	87.77	0.30	0.29	0.225	1.219
8	300	300	1.82	2.32	73	64	72.62	92.69	0.25	0.31	3.213	0.738
9	300	400	2.48	2.78	103	90	99.25	111.34	0.33	0.38	1.581	-[a]

[a] The destruction of specimen after t=5 min.

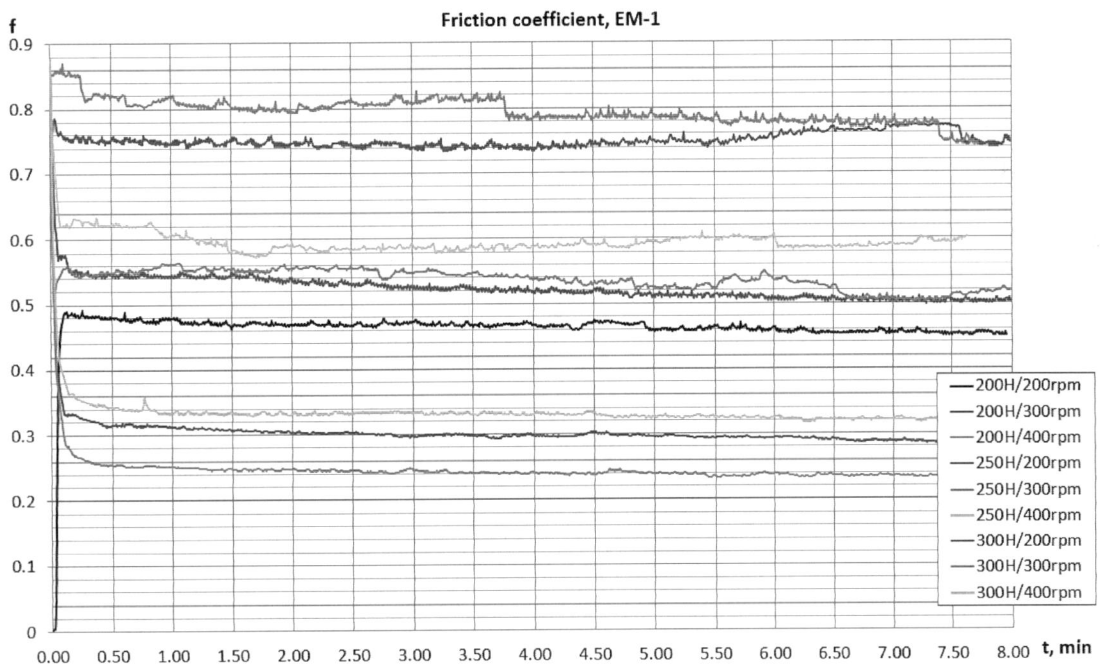

Figure 2. Friction coefficients of EM-1 material.

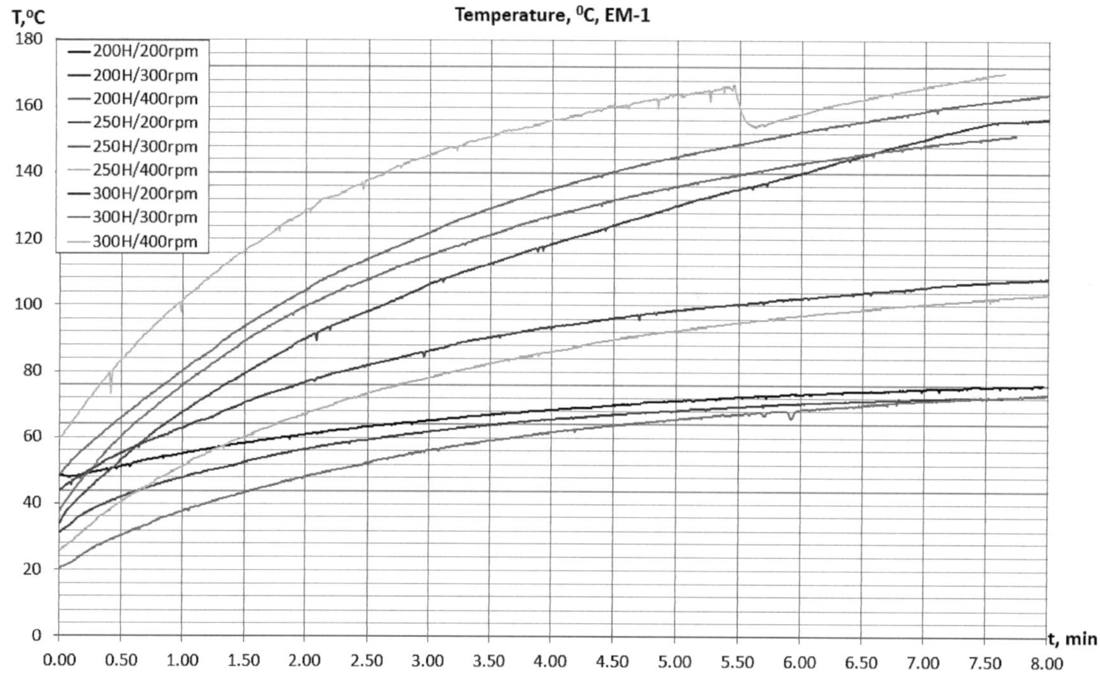

Figure 3. Temperatures of EM-1 material.

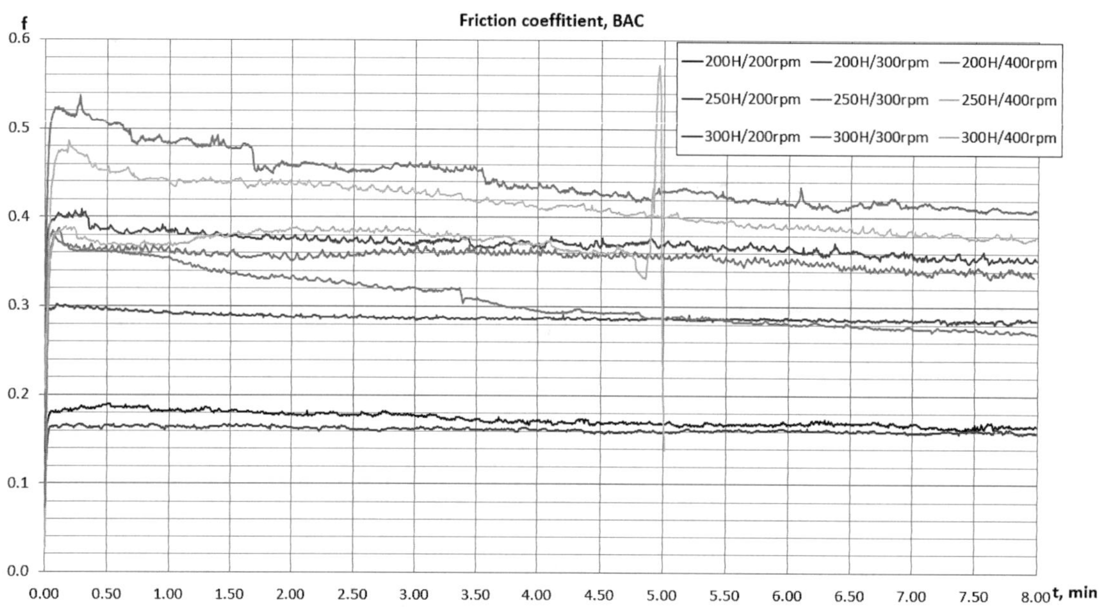

Figure 4. Friction coefficients of BAC material.

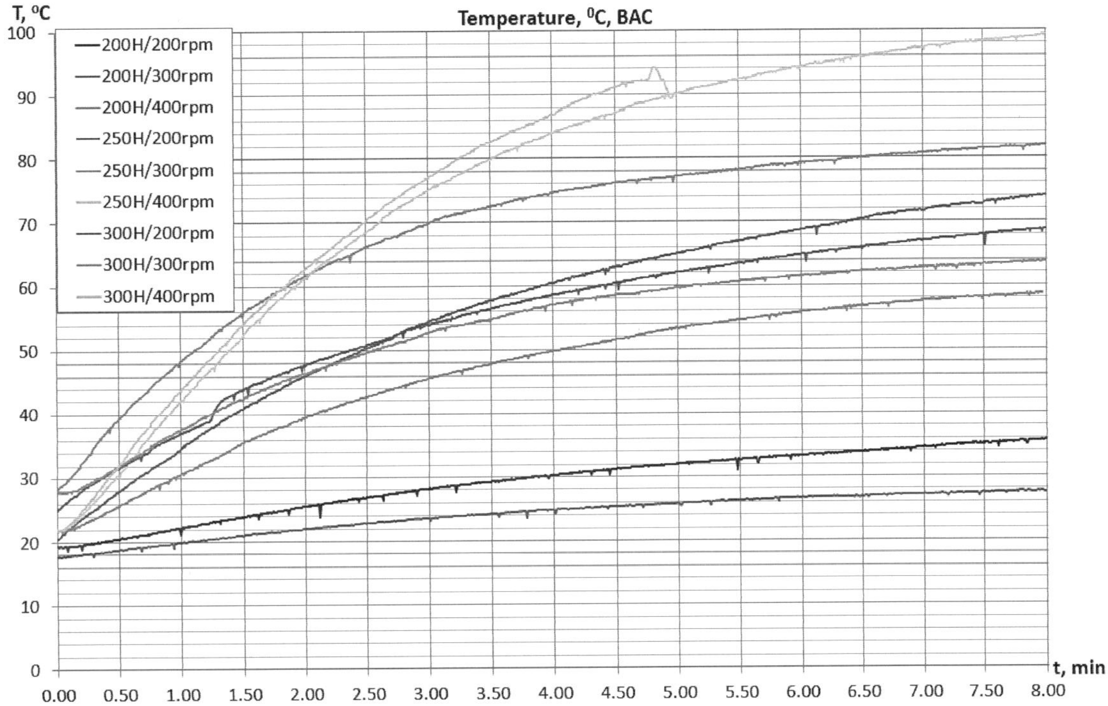

Figure 5. Temperatures of BAC material.

4. Discussion

The experimental data analysis showed different specimen behavior under nominal loads. In average the EM-1 material heats up to higher temperatures yet provides higher friction coefficients. In mode №1 the final temperature and average friction coefficients are pretty much the same. The exaggeration of test conditions gives ambiguous results. So, increasing the rotational velocity at 200 N load results in really high friction coefficient of 0.8 at temperature of 150°C. This friction coefficient is twice higher in comparison to BAC band. The final temperature is 2.5 times higher in comparison to BAC band. In modes 5-6 (which are harder than modes 2-3) drastically lower friction coefficients were observed while the temperatures were higher. After exceeding the nominal unit pressures in modes 7-9 the friction coefficient turned out to be remarkably low, 0.25-0.33 while the temperature varied from 73°C to 103°C. There values are significantly lower than nominal values, and the temperatures are lower in comparison to previous modes.

The BAC band performance is much more stable. So, the friction coefficients under nominal loads were 0.4±10%. It must be noted that in modes 2-3 the EM-1 band provided the highest friction coefficients while the BAC band – the lowest. In most modes the temperature of BAC band was 1.5-2 times lower in comparison to EM-1 band. In modes 7-8 the performance similar to EM-1 band was noted: the average friction coefficients were lower than nominal, the specimen temperatures were quite low. During the test in mode 9 (300 N, 400 rpm) the friction coefficient was approximately 0.38 at temperature rise up to 90°C. At the end of the fifth minute of experiment the destruction (interlaminar shear) of the specimen took place.

Figure 6. The post-test sliding surface of EM-1 material.

Figure 7. The post-test sliding surface of BAC material.

The average weight wear following 8 tests was 1.56 mg·min^{-1} for EM-1 and 0.88 mg·min^{-1} for BAC bands. Tests №9 were not included due to destruction of BAC №9 specimen with significant loss of material. Visual estimate also showed much higher wear levels of EM-1 material. Also, the edge chipping of EM-1 material was noted. The images of specimen sliding surfaces are shown in figures 6-7.

5. Conclusions

The comparative tests have shown that while operating the brake systems in nominal modes asbestos-containing EM-1 material provides higher braking torque at the price of lower wear resistance, higher operating temperatures and higher ecological damage. The scattered values of friction coefficients in different operating modes were noted. The BAC bands performance is much more predictable while providing stable nominal friction coefficients with scatter of ±10% and lower operating temperatures, as well as lower weight wear values. The performance of both materials under excess load is unsatisfactory with friction coefficients being lower than nominal. The BAC bands provide stable braking parameters at the same time proving 1.7 times higher wear resistance, thus usage of these bands is seen more economically viable: the band replacement frequency is lower, as well as the maintenance of brake systems. Lower average operating temperatures may result in increasing the life cycle of lubricants in the units of brake systems.

References

[1] Sinelnikov A F 2008 *Management of maintenance and rework of lifting, transport, construction and road-mending machinery under different operational conditions* (Moscow: Academia Publishing) 336 p

[2] World Health Organization URL: https://www.who. int/ipcs/assessment/public_health /asbestos/en/

[3] Kulakov O I and Albagachiev A Ju 2019 The modernization of friction testing machine II 5018 *Collection of conference papers of VI int. scientific conf. FRITME-2019* (Moscow: IMASH RAS) pp 244-245

IOP Conf. Series: Materials Science and Engineering **996** (2020) 012017 doi:10.1088/1757-899X/996/1/012017

Features of wear and friction in titanium

M A Pakhomov[1], D Gorlov[2] and V Stolyarov[1]

[1] Mechanical Engineering Research Institute of the Russian Academy of Sciences, Moscow, Russia
[2] All-Russian Scientific Research Institute of Aviation Materials, Moscow, Russia

pakhomovmish@gmail.com, gorlov80@yandex.ru, vlstol@mail.ru

Abstract. The characteristics of friction and wear during low-frequency reciprocating motion of the indenter in commercially pure titanium with coarse and fine grains are investigated. The role of the microstructure refining and the ratio of grain size and displacement amplitude is shown.

1. Introduction

Fretting wear is considered a special case of friction between two contact surfaces at very low amplitude <0.5 μm and a frequency of 0.001 Hz of cyclic movements [1]. During fretting, the contact surfaces are not removed from the contact area, and, therefore, the fracture products have almost no possibility to get out of the wear zone [2]. Fretting wear results in fretting fatigue, which leads to a decrease in fatigue strength and the number of cycles to failure by tens of percent [3, 4]. The scientific significance of research on fretting wear is associated with an understanding of the nature and mechanisms of friction in conditions of low amplitudes and frequencies of micromovements, as well as the possibility of modeling the wear process. In practical terms, the study of fretting wear is important for the attachment points of the compressor blades of aircraft engines, fuel elements in a nuclear reactor, riveted and bolted joints, bearings, electrical contacts, dental implants and other similar structures. Nevertheless, most of the tribological studies are still carried out in sliding friction mode. Titanium alloys, due to their high performance properties, are often used for the similar designs mentioned above. However, insufficiently high strength, poor thermal conductivity and high friction coefficient C_f lead to their high sensitivity to fretting fatigue. Therefore, studies of fretting wear of titanium alloys were very relevant, however, they were carried out for a material in a traditional coarse-grained state [5-8]. Among the many factors influencing fretting wear, displacement amplitude, test temperature, and grain size seem to be the most important in the study of friction and wear in structural materials. Unfortunately, there are relatively few works devoted to studying the effect of structure refinement [9-11]. Significant progress in understanding the role of structural refinement has been achieved in the study of the friction of ultrafine-grained (UFG) titanium alloys [12-14]. Note that the above studies of UFG and nanostructured (NS) titanium alloys were carried out under traditional friction conditions, far from fretting. As for the role of temperature in fretting, the study of two-phase coarse-grained titanium Ti-Sn-Al-Mo-Zr alloys in the temperature range from 20 to 600°C showed that the formation of oxide films on the surface of the samples contributes to a decrease in the C_f and the wear rate with an increase in temperature [1]. The purpose of this work was to study the effect of grain size on the fretting behavior of commercially pure titanium at room temperature and a relatively small amplitude of indenter displacement.

2. Material and research methods

The starting material was commercially pure titanium Grade 4 in the form of a bar 20 mm in UFG state (grain size $d_z = 0.45$ μm), obtained by equal channel angular pressing. Two plates 2 x 6 x 43 mm^3 were cut from the bar along the longitudinal axis, one of which was subjected to annealing (680°C, 1 hour) to obtain a coarse-grained (CG) state (d_z = 45 μm). To prevent the possible influence of roughness, the sample plates were polished before testing for fretting wear so that the surface roughness would be Ra ≤ 0.15 μm. Fretting wear tests were carried out without lubrication on a CETR UMT-3M tribometer using a reciprocating motion in a direction parallel to the pressing axis of the sample with a displacement amplitude of A = 300 μm and a frequency of 13 Hz (3.9 $mm \cdot s^{-1}$). The experiment was carried out in air, at a room temperature of 20°C, for 128 min. Tungsten carbide ball of ∅6 mm as counter body was used, the load was 3 N. The choice of the force value is based on the capabilities of the device (on the one hand) and the idea of the independence of C_f on the normal force [15]. The total linear wear of the sample and the indenter was recorded by a stepping motor. For greater reproducibility of the wear test results for each mode, measurements were made at three points. The morphology of the wear places was examined using an Olympus SZ microscope. Fretting behavior was assessed by the dependences of wear and C_f on time, as well as by the wear morphology of contact points.

3. Results

Figure 1 shows the microstructures of UFG titanium (figure 1 (a)) and CG titanium (figure 1 (b)).

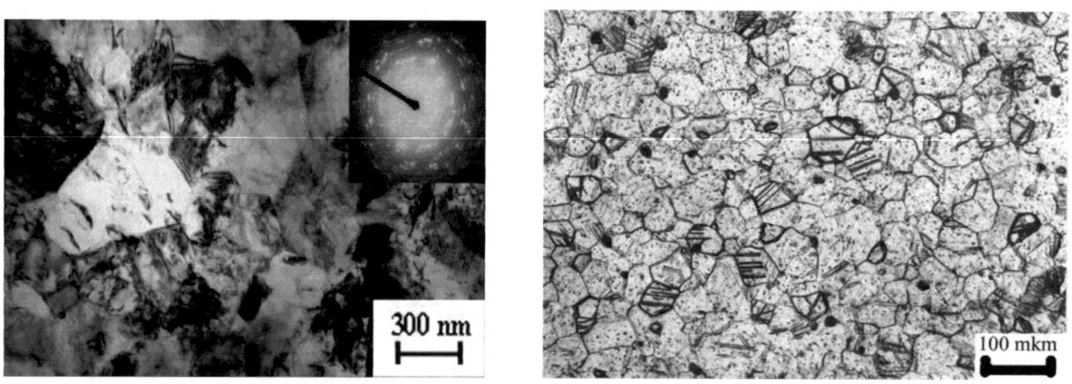

(a) (b)

Figure 1. Microstructures of ECAP titanium before (a) and after (b) high temperature annealing.

Figure 2 shows typical dependences of wear and C_f on time for CG titanium at room temperature. It is noted that, with the exception of the initial stage, the wear is positive and increases exponentially with time (figure 2 (a)). In this case, C_f has a wave-like dependence in time and high values in the range from 0.8 to 1.2 (figure 2 (b)).

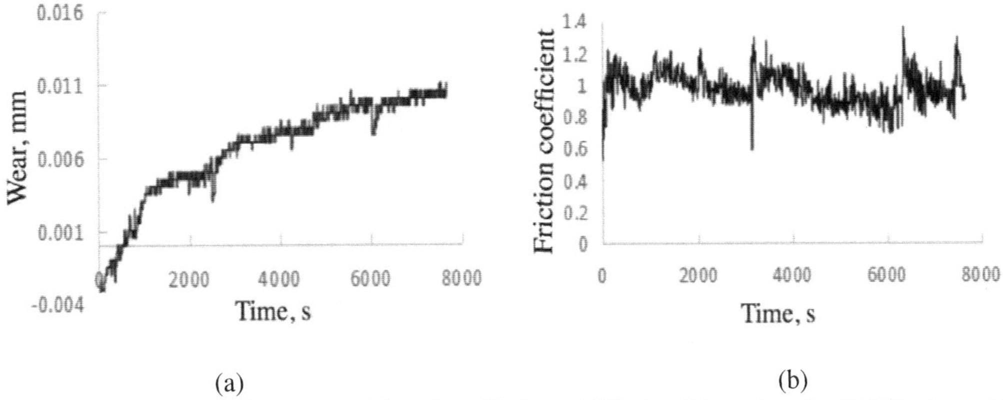

Figure 2. Dependence of linear wear (a) and coefficient of friction (b) on time for SC titanium at 20°C and A = 300 μm.

Figure 3 shows typical dependences of wear and C_f on time for UFG titanium at room temperature and amplitude of 300 microns. For UFG titanium, wear over time is characterized by a decrease in the initial wear rate to 5 μm·hour^{-1} (figure 3 (a)). The C_f curve has stages of decrease, reaching a stationary stage (plateau), followed by its increase (figure 3 (b)).

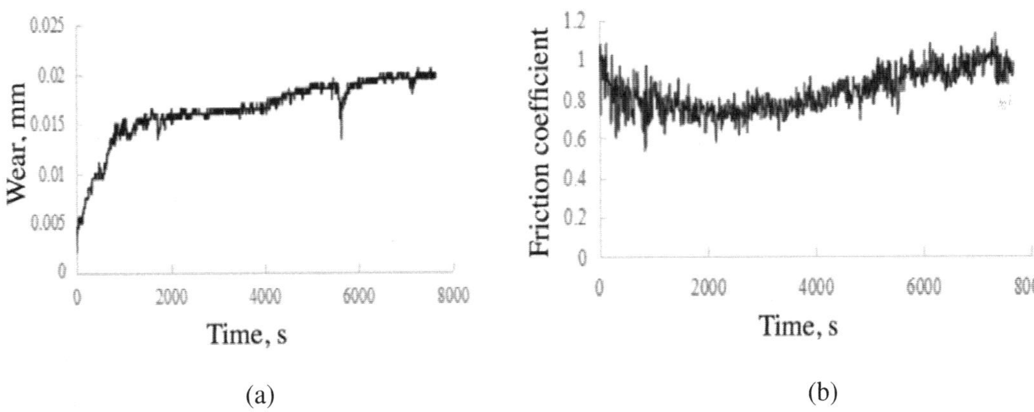

Figure 3. Dependence of linear wear (a) and friction coefficient (b) on time for UFG titanium at 20°C and A = 300 μm.

Table 1 summarizes the results of measuring the average values of wear, C_f, the initial size of the contact spot for CG and UFG titanium at a temperature of 20 °C and a displacement amplitude of 300 μm at the end of the tests. Maximum wear has positive values and corresponds to UFG titanium. At displacement amplitude of 300 μm, this value is twice that for CG titanium.Structure refinement did not affect the friction coefficient, the maximum value of which is 1.

The mechanism of friction and its features in UFG and KZ titanium, the morphology of contact spots of the samples with the indenter were studied using an optical microscope. The general view of the contact spots after tests at room temperature is shown in figure 4.

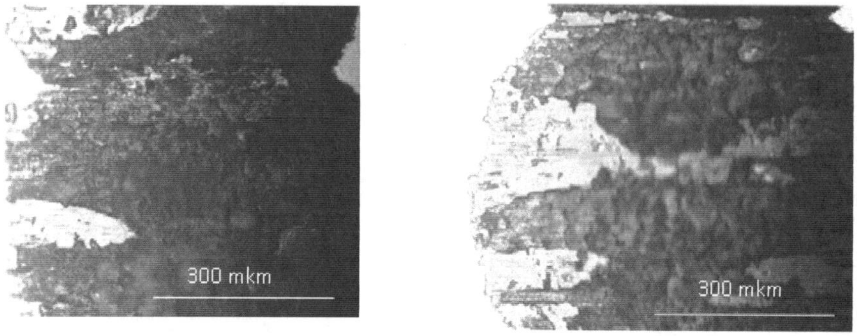

(a) (b)

Figure 4. General view of the contact spots of CG (a) and UFG (b) titanium after testing at 20°C and A = 300 μm.

It can be seen that the morphology of contact spots for CG and UFG titanium is typical for abrasive wear and practically does not depend on the structure refinement. The linear size of the contact spots at the end of the test exceeds the displacement amplitude by 3 and 2.5 times, respectively. The adhesion zone typical for fretting wear, as well as wear products around the contact patch, are absent.

Table 1. Wear and friction coefficient.

Structure state	Hv, (MPa)	Initial contact area size (R, μm)	Displacement Amplitude (A, μm)	Wear (μm)	C_f
CG	2860	~75	300	10	1.0
UFG	3320	~70		20	1.0

4. Discussion

The results obtained showed a number of features in the tribological behavior of titanium. In both structural states, wear develops over time in two stages, which differ in the wear rate - accelerated at the beginning and slowed down at the end. The deceleration of the wear rate and its stabilization over time indicates the formation of a "third body" between the contacting surfaces.

Despite the higher microhardness of UFG titanium, the total amount of wear in this state is twice that for CG titanium (table 1). This is most likely due to the density of grain boundaries, which in UFG titanium is several orders of magnitude higher than in CG titanium. Since it is known that grain boundaries are a source of defects and places of destruction, their multiple intersections lead to increased wear.

The different sign of the wear in the CG and UFG titanium at the initial stage of testing is due to the different thickness of the oxide film and its subsequent destruction [1]. It is known that the corrosion resistance of UFG titanium is higher than the similar characteristic of CG titanium, which is associated with the high rate of formation of a protective film with a more dispersed structure [17]. The average C_f values for both states were approximately the same and high. A distinctive feature is the presence of wavelike behavior of the time dependence of the friction coefficient for CG titanium (figure 2 (b)). One of the explanations for this phenomenon can be the processes of hardening and softening due to refinement and destruction of the surface layer ("third body") between the indenter and the material. This process also manifests itself for UFG titanium, but with a long period (figure 3 (b)).

The ratio of the displacement amplitude and the size of the contact zone A/R> 1 shows that the conditions for fretting are not realized and the wear process corresponds to traditional friction [2].

Noteworthy is the many times higher size of the contact spots in comparison with the displacement amplitude. Perhaps this fact is associated with an increase in temperature on the contact surface during the movement of the indenter.

5. Conclusions

1. The tribological behavior of titanium is a complex phenomenon that depends not only on the hardness and corrosion resistance of the surface layer, but also on the structure refinement.
2. Wear in UFG titanium is higher than in CG titanium, while the friction coefficient does not depend on the grain size.
3. The investigated conditions of the experiment, the amplitude and frequency of movement, exclude the possibility of fretting wear.

References

[1] Waterhouse R 1984 Fretting wear *Wear* **100** 107-18
[2] Waterhouse R 1972 Fretting Corrosion *Pergamon* **26** 133
[3] Mall S, Kim K, Saladin C and Porter J 2010 Effects of microstructure on fretting fatigue behavior of IN100 *Materials Science and Engineering* **527** pp 1453-60
[4] Nowell D, Dini D and Hills A 2006 Recent developments in the understanding of fretting fatigue *Engineering Fracture Mechanics* **73** pp 207-22
[5] Zhang H, Liu X, Liu H, Wang Y and Tang B 2011 Improvement of the fretting damage resistance of Ti-811 alloy by CuNi multilayer film *Tribology Int.* **44** pp 1488-94
[6] Vadiraj A and Kamaraj M 2007 Effect of surface treatments on fretting fatigue damage of biomedical titanium alloys *Tribology Int.* **40** pp 82-88
[7] Fridrici V, Fouvry S and Kapsa P 2001 Fretting Wear Behavior of a Titanium Alloy *Springer Dordrecht* **10** pp 413-21
[8] Blanchard P, Colombie C, Pellerin V, Fayeulle S and Vincent L 1991 Material effects in fretting wear: application to iron, titanium, and aluminum alloys *Metallurgical and Materials Transactions* **22** pp 1535-44
[9] Zhang S and Han Z 2007 Fretting wear behavior of nanocrystalline surface layer of pure copper under oil lubrication *Tribology Letters* **1** pp 53-59
[10] Wang T, Gao N, Mark G, Gee, Wood K and Terence G 2013 Tribology testing of ultrafine-grained Ti processed by high-pressure torsion with subsequent coating *J. of Materials Science* **48** pp 4742–48
[11] Misochenko A, Chertovskikh V, Shuster S and Stolyarov V 2017 Influence of Grain Size and Contact Temperature on the Tribological Behavior of Shape Memory Ti49.3Ni50.7 Alloy *Tribology Letters* **4** 131
[12] Fu Y, Bellam C, Qazi I, Rack J and Stolyarov V 2004 Reciprocating-sliding wear of ultra-fine grained TI-6AL-4V *Ultrafine Grained Materials* **III** pp 547-52
[13] Migranov Sh, Shuster Sh and Stolyarov V 2004 Tribological properties of ultrafine-grained titanium obtained by the method of severe plastic deformation *Mechanical engineering bull.* **7** 37
[14] Rabinowicz E 1995 Friction and wear of materials *Wiley 2nd ed* 315
[15] Popov L 2013 Mechanics of contact interaction and physics of friction. From nanotribology to earthquake dynamics *Fizmatlit* 352
[16] Yu Y, Quan X, Cai B, Gao S and Zhu H 2008 Radial Fretting Behavior of Cortical Bone Against Titanium *Tribology Letters* **69** pp 69-76
[17] Balyanov A *et al* 2004 Corrosion resistance of ultra fine-grained Ti *Scripta Materialia* **3** pp 225-29

IOP Conf. Series: Materials Science and Engineering **996** (2020) 012018 doi:10.1088/1757-899X/996/1/012018

Tribological activity of ultradisperse organic lubricant additives for cam mechanisms subject to abrasive wear

A S Parfenov[1], E V Berezina[1], I V Berezin[2,3], A V Volkov[4] and V A Godlevsky[5]

[1] Ivanovo State Medical Academy, 8 Sheremetev Avenue, Ivanovo 153012, Russia
[2] National Research University Higher School of Economics, 20 Myasnistskaya Str, Moscow 101000, Russia
[3] D. Mendeleev University of Chemical Technology of Russia, 9 Miusskaya sq, Moscow 125047, Russia
[4] MBOU Gymnasium № 44, 40 Kohom Str, Ivanovo 153051, Russia
[5] Ivanovo State University, 37/7 Ermaka Str, Ivanovo 153025, Russia

alsparf@gmail.com

Abstract. The effect of nanodispersed triboactive additive introducing into the oil suspension of model abrasive is considered. The influence of the concentrations both additive and abrasive, as well as the dispersion of the solid phase on the wear of steel-steel pair friction, is determined. It has been suggested that carbon nano-particle additives under these conditions can shield the level of abrasive wear.

1. Introduction

Cam mechanisms which are suffering from abrasive wear and those mechanisms are widely used in automatic machines of textile, publishing and light industry.

A change in the geometry of the profile of a cam mechanism due to wear leads to blows (changes in acceleration) that destroy the contact surfaces of the cam and the pusher. In many respects, the wear resistance of the cam mechanism depends on the value of contact stresses, the quantitative characteristic in the contact is the calculation of the contact strength of cams and rollers according to the Hertz-Belyaev formula. In this regard, it is important to take into account the contamination of lubricant with abrasive particles.

Many studies on the behavior of abrasive particles in the lubricant are devoted to (see, for example, [1-3]). They consider abrasive wear models under conditions of liquid-flowing, plastic lubricants or without lubrication. In recent years, the study of nanodisperse additives in tribology as systems with high specific surface area has been actively developed. The application of additives based on carbon nanoparticles (in the form of allotropic carbon modifications: single-wall and multi-wall carbon nanotubes, graphene and graphene oxide, schungite nanoparticles) has attracted increased interest in recent decades [4-8]. Multi-Wall Carbon Nanotubes (MWCN), which is an allotropic modification of carbon that has a cylindrical structure with a diameter of up to tens of nanometers and length from 1 μm to several centimeters, is of interest among researchers [8-13] due to the ability to improve tribological characteristics of lubricants and to form a film on the friction surfaces that reduces the wear and tear of parts. An important characteristic that distinguishes MWCN from other carbon nanomaterials is their high hardness [14].

Purpose of following research is:

- to check influence of MWCN as systems with higher hardness among all other carbon allotropic modifications;
- to find connection of its molecular structure with non-abrasive action.

2. Materials and methods

As an oil material was used oil I-40A (Russian standart 20799-88, analogue of BP Energol CS 32).

As an active abrasive agent of oil material was injected abrasive material- Electrocorundum 25A 0.063-0.125 (F180) (Russian standart28818—90,analogue of White fused aluminium oxide(corundum) EK). Main characteristics of Corundum are in table1 and figure 1.

Table 1. Main characteristics of Corundum 25A.

Characteristic	Corundum 25A (F180)
Fraction composition, nm	0.063-0.125
Main size of fraction, nm	0.075-0.090
Content of Al_2O_3	96-99%
Microhardness, GPa	19.6-20.9
Density, $g \cdot cm^{-3}$	3.9-3.95

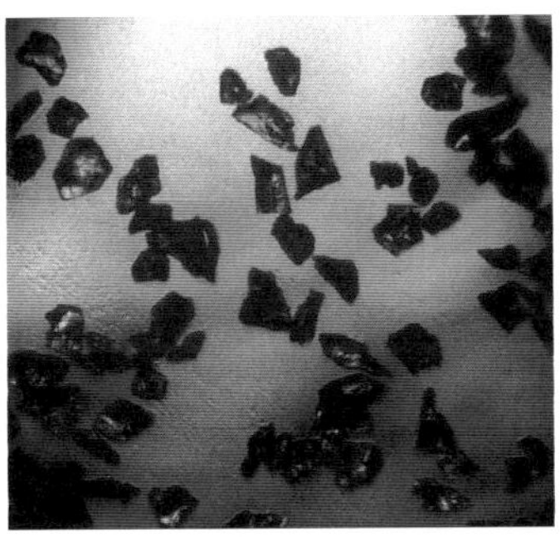

Figure 1. Characteristic shape of abrasive corundum particles.

As an additive is was used carbon nanotubes Taunit-M (produced by «NanoTechCenter Co., LTD», Tambov city).Carbon nanotubes (UNT) line «Taunit» - quasi one-dimensional, nanoscale, filamentous formations of polycrystalline graphite with mostly cylinder form with inner channel (table 2).

Table 2. Characteristics of MWNT Taunit-M.

Characteristic	Taunit-M
Internal diameter, nm	10-30
Internal diameter, nm	5-15
Length, μm	≥2
Specific surface, $m^2 \cdot g^{-1}$	≥270

Abrasive agent and UNT additive were injected into Oil Material (OM) by mechanical mixing and followed ultrasonic dispersion. Tribologic tests were made at universal friction machine MTU-01 (TS 4271-001-29034600-2004). Rotational speed of shaft kept constant – 200 rpm. As a friction pair were used 3 balls with diameter 12.7 mm, steel ShKh15 – analogue of AISI Steel 52100, hardness 60-62

IOP Conf. Series: Materials Science and Engineering **996** (2020) 012018 doi:10.1088/1757-899X/996/1/012018

HRC (Russian Standart 3722-2014)and steel plate, steel R6M5 - analogue of AISI M2 HSS (Ra 1.25 μm, hardness 65 HRC). Scheme of the device is on figure 2.

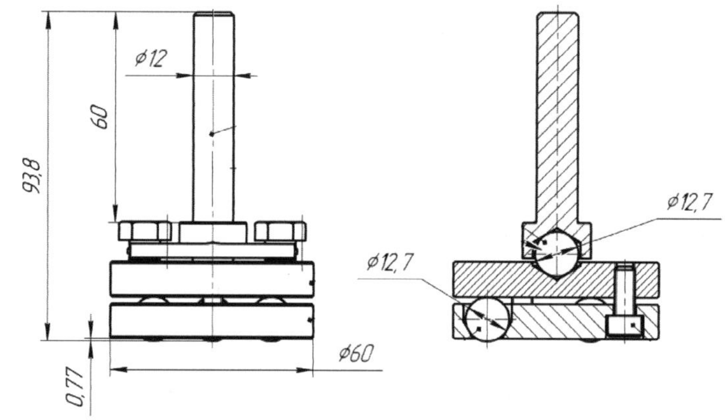

Figure 2. Geometric parameters of the friction pair.

The wear scar diameter measured on a microscope Biolam-M with digital camera Levenhuk M500 Base, using the software ToupView.

3. Results of the experiment and their discussion

3.1. Theoretical model

During the process of abrasive wear of shaft, it will be checked geometry of contact. Let's consider following situation.

S – is an area of contact, n –number of particles of abrasive agent, effected to this area, $l = \upsilon \cdot t$ – glide way, passed during time t with velocity υ. Surface wear will be both from cam and from pusher roller. We accept, that area is function of time $S = S(t)$.

If we talk about cam mechanism of textile machine (e.g, in sealed box), vertical force F will be changed according to the periodic law

$$F = b\sin^2 \omega t,$$

Where b –coefficient of proportionality, ω –frequency of interaction in the sliding friction mode. Constant b and frequency ω are determined based on the geometry of the friction pair.

Let's enter: $P(S_n)$ – probability, that with $P_0 = \frac{F}{S}$ pressure and velocity υ a wear particle will appear.

Thus, the number of wear particles N_w will increase over time

$$N_w(t) = PN_k(t) \tag{1}$$

N_k – is a number of abrasive material additives during the way l.

In case if, cam mechanism wear will be more, than pusher roller wear, number of abrasive particles n_k, per unit run length, that

$$N_k(t) = n_k l = n_k \upsilon t \tag{2}$$

And the number of wear particles

$$N_w(t) = Pn_k \upsilon t \tag{3}$$

In this case, the assumed probability model is

$$P = e^{-\alpha \frac{A_n}{A_\tau}} \quad (4)$$

where α -the ratio of the material hardness to the hardness of the abrasive particles, what characterizes their cutting ability, A_n - work of normal forces, A_τ - work of tangential forces, which depends on the parameters of the lubricant oil, temperature, abrasive particles geometry.

In this case, wear rate I_w of the mating part of the part (cam mechanism) - $I_w = \frac{dN_k(t)}{dt} = Pn_k\upsilon$ equilibrium value of surface wear at a given speed. Wear particles have a shielding effect on the entrance and exit of the friction zone of corundum particles, fixing the concentration.

When both parts of the friction pair wear out

$$n_k = n_k(t) \quad (5)$$

Then

$$N_w^i = P^i n_k(t)\upsilon t \quad (6)$$

And wear speed

$$I_w^i = \frac{dN_w^i(t)}{dt} = P^i \upsilon \frac{dn_k(t)\cdot t}{dt} \quad (7)$$

When using the model in practice, to work correctly, you need to define $n_k(t)$ for a specific friction unit, which depends on the geometry of the friction pair (in particular, the cam mechanism).Then the dependence of the number of wear particles on time takes the form:

$$n_k(t) = \frac{\alpha}{2}t + n_k(0) \quad (8)$$

In this case, the wear rate is set by the expression.

$$I_w^i = \frac{dN_w^i(t)}{dt} = P^i \upsilon \frac{d}{dt}\left(\frac{\alpha}{2}t^2 + n_k(0)t\right) = P^i \upsilon (n_k(0) + \alpha t) \quad (9)$$

If additives dispersed to nanoscale are additionally introduced into the system, which are able to bind to abrasive particles, they will, first of all, have a screening effect on the particles of the abrasive material, thereby reducing their cutting effect, and, consequently, reducing the ratio in the constant of compliance to wearα.

In the process of abrasive wear, in addition to the initial particles of the abrasive material, wear particles of the friction surfaces will also be formed, which are metal oxides that have an increased hardness relative to the friction surfaces. Such a mixture can be considered as a polydisperse system consisting of abrasive material particles, wear particles formed during friction and additionally introduced ultradisperse particles introduced as an additive.

Wear particles can play an important function when introducing an ultrafine additive. When they are located in the friction zone, the ultrafine additive is transported both on the surface of the abrasive particles and with the help of wear particles. During the friction process, in the case of additives, it is possible to stabilize the system state and bring it to an equilibrium state, which avoids a catastrophic increase in the rate of abrasive wear due to an avalanche-like increase in the number of wear particles.

3.2. Research of I40-Taunit-M systems

At the first stage of the study, the effect of the introduction of the additive "Taunit-M" was determined at OMI-40A. The introduction of the additive led to a decrease in the diameter of the wear spot at concentrations from 0.5 up to 2.0 weight%. At the same time, the composition showed the greatest effect with 1.5 wt. % (figure3)

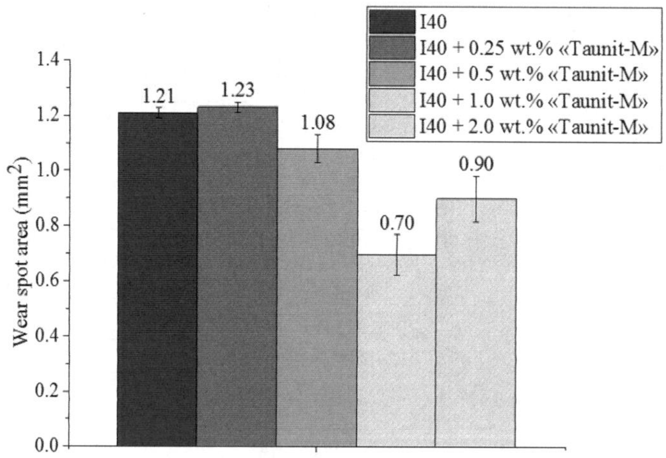

Figure 3. The wear spot area.

At the same time, the introduction of an additive at all concentrations led to an increase in the friction moment (figure 4)

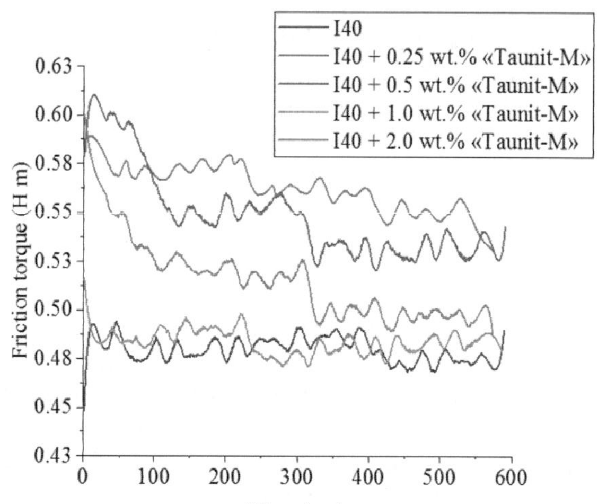

Figure 4. The values of the friction torque during the test.

The obtained photos of wear spots may indicate that when carbon nanotubes are introduced as a lubricant additive, their effect largely consists in the formation of a carbon film on the contacting surfaces. As you can see (figure 5), with an increase in the concentration of the additive, a visual increase in carbon is noticeable on the surface of the wear spot, which may indicate work in the process of friction of carbon films.

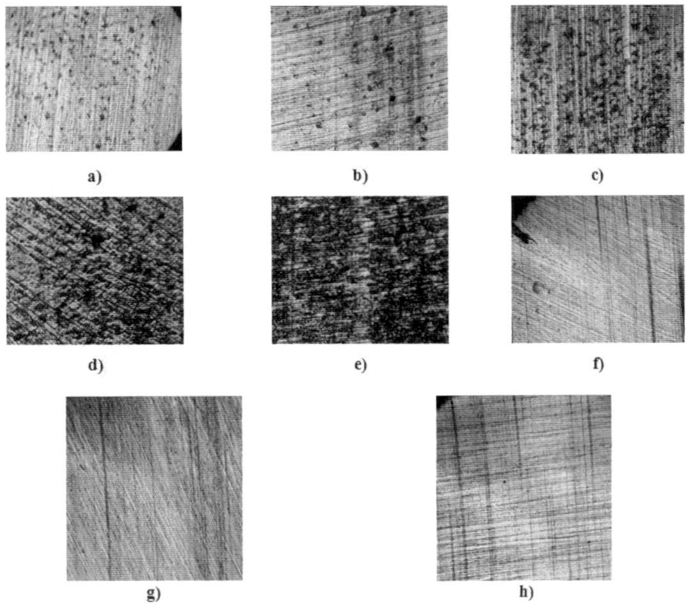

Figure 5. Structure of the wear spots: (a) I40; (b) I40 + 0.25 wt. % «Taunit-M»; (c) I40 + 0.5 wt. % «Taunit-M»; (d) I40 + 1.0 wt. % «Taunit-M; (e) I40 + 2.0 wt. % «Taunit-M»; (f) I40 + 0.5 wt. % Corundum; (g) I40 + 0.5 wt. % Corundum + 1.0 wt. % «Taunit-M»; (h) I40 + 0.5 wt. % Corundum + 2.0 wt. % «Taunit-M».

3.3. Research of I40-Taunit-M systems contaminated with abrasive material

At the second stage of the study an abrasive material Corundum was introduced into the lubricant 25A 0.063-0.125 (F180).

Tests have shown that adding of 0.5 wt. % abrasive agent leads to a sharp increase in the diameter of the wear spot (more than twice). While this, the introduction of "Taunit-M" led to a decrease in the abrasive effect of corundum particles.

The greatest effect from the introduction of nanoparticles was 23% in case of mass concentration of the additive 1.0 wt. % (figure6).

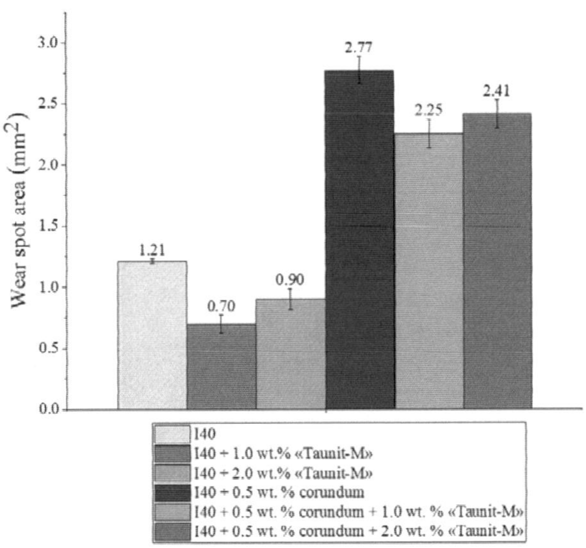

Figure 6. Area of the wear spot when introducing abrasive material.

At the same time, when the abrasive material was introduced, the friction mode was unstable. In the range from 250 to 350 seconds after the start of the test, there was a sharp increase in the friction moment. In the case of a system with a lubricant contaminated with abrasive, an abrupt increase in the friction moment can be observed, which may indicate instability. Introduction of the additive at a concentration of 1.0 wt. % had no effect on this phenomenon, but an increase in the concentration to 2.0 wt. % resulted in stabilization of the (figure7).

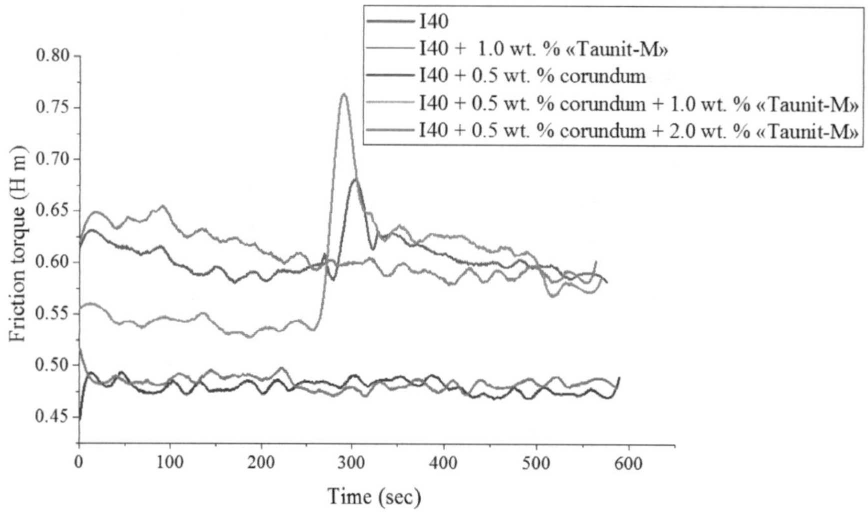

Figure 7. Area of the wear spot when introducing abrasive material.

The dependence of the area of the wear spot on the additives and abrasives introduced can be represented as a histogram (figure 6).

The temperature dependence was also analyzed during the experiment. Introduction of carbon nanotubes in concentrations of 0.25 and 0.5 wt. % resulted in a slight increase in temperature at the end of the test. The introduction of UNT Taunit-M at a concentration of 1.0 and 2.0 wt. % resulted in a slight decrease in system temperature.

In the case when the I40 lubricant was contaminated with abrasive corundum particles, the additional introduction of carbon nanotubes led to an increase in the system temperature (table 3).

Table 3. Temperature variation during testing.

Lubrication material	Temperature variation, °C
I40	13.79
I40 + 0.25 wt. % «Taunit-M»	14.21
I40 + 0.5 wt. % «Taunit-M»	13.26
I40 + 1.0 wt. % «Taunit-M»	11.06
I40 + 2.0 wt. % «Taunit-M»	11.34
I40 + 0.5 wt. % Corundum 25A	15.08
I40 + 0.5 wt. % Corundum 25A + 1.0 wt. % «Taunit-M»	15.61
I40 + 0.5 wt. % Corundum 25A + 2.0 wt. % «Taunit-M»	18.06

4. Conclusion

As a result of tests, it was shown that the introduction of carbon nanostructures "Taunit-M" as an additive reduces wear both in the base oil and in the Oil with the addition of abrasive particles. Photos of the friction surface after testing may indicate that in the case of base oil, the reduction in wear is due to the formation of an additional layer of graphite between the contacting surfaces.

In the case of a composition with abrasive particles, there are practically no traces of carbon on the friction surface. In this case, we can consider the interaction of carbon nanotubes directly with abrasive particles. Abrasive wear can be reduced due to the "shielding" effect of nanoparticles, which, when they come into contact with corundum, reduce the cutting effect of particles on the surface of the ball.

It can be assumed that carbon nanotubes can shield the cutting action of abrasive particles by electrostatically interacting with them. Carbon nanotubes, having a sufficiently large stiffness and small size, are fixed to the abrasive particles, damping the impact of the abrasive particle on the friction surface, while passing directly to the surface itself. To sum up, two processes can occur: reduced cutting action of the abrasive and additional transfer of carbon nanoparticles to the friction surfaces.

References

[1] Ramadan M A 2018 Friction and wear of sand-contaminated lubricated sliding *Friction* **6** 457–63

[2] Salehiandastjerdi M, Esehaghbeygi A, Razavi J and Ghaziasgar H S 2019 Effect of calcite and silica content on tribological performance of engine oil lubricants *Tribol. Online* **14** 1–7

[3] Sviridenok A I, Ihnatouski M I, Smurugov V A, Chmykhova T G and Khovatov P A 2012 Tribological properties of lubricants modified by complexes of hard micro- and nanoparticles *J. Frict. Wear* **33** 266–73

[4] Kim H J, Yoo S S and Kim D E 2012 Nano-scale wear: A review *Int. J. Precis. Eng. Manuf.* **13** 1709–18

[5] Lee K, Hwang Y, Cheong S, Choi Y, Kwon L, Lee J and Kim S H 2009 Understanding the role of nanoparticles in nano-oil lubrication *Tribol. Lett.* **35** 127–31

[6] Johns E I, Padgurskas J, Zunda A, Kupcinskas A and Kavaliova I N 2018 Study of Tribological Properties of Tallow Modified by Graphene and Expanded Graphite *J. Frict. Wear* **39** 341–4

[7] Voronin S V, Suranov A V and Suranov A A 2017 The effect of carbon nanoadditives on the tribological properties of industrial oils *J. Frict. Wear* **38** 359–63

[8] Zin V, Agresti F, Barison S, Colla L, Mercadelli E, Fabrizio M and Pagura C 2014 Tribological properties of engine oil with carbon nano-horns as nano-additives *Tribol. Lett.* **55** 45–53

[9] Usol'tseva N V, Smirnova M V, Kazak A V, Smirnova A I, Bumbina N V, Ilyin S O and Rozhkova N N 2015 Rheological characteristics of different carbon nanoparticles in cholesteric mesogen dispersions as lubricant coolant additives *J. Frict. Wear* **36** 380–5

[10] Zhang Z, Simionesie D and Schaschke C 2014 Graphite and Hybrid Nanomaterials as Lubricant Additives *Lubricants* **2** 44–65

[11] Cursaru D L, Ramadan I, Tanasescu C and Ripeanu R 2013 Study of the tribological behavior of different carbonaceous nanomaterials such as antiwear additives for an environmentally friendly lubricant *Dig. J. Nanomater. Biostructures* **8** 805–15

[12] Khalil W, Mohamed A, Bayoumi M and Osman T A 2016 Tribological properties of dispersed carbon nanotubes in lubricant *Fullerenes Nanotub. Carbon Nanostructures* **24** 479–85

[13] Shahnazar S, Bagheri S and Abd Hamid S B 2016 Enhancing lubricant properties by nanoparticle additives *Int. J. Hydrogen Energy* **41** 3153–70

[14] Shilov M A, Smirnova A I, Stolbov D N and Usol'tseva N V 2020 Modelling of deformation processes of carbon nanotubes *Liq. Cryst. and their Appl.* **20** 85–91

IOP Conf. Series: Materials Science and Engineering **996** (2020) 012019 doi:10.1088/1757-899X/996/1/012019

On the prospects of research in the field of evaluation and reduction of mining machines elements wear

P A Pobegailo[1], I V Gadolina[1] and Ju A Gustov[2]

[1] Mechanical Engineering Research Institute of the Russian Academy of Sciences (IMASH RAN), Moscow, Russia
[2] Moscow State University of Civil Engineering (MGSU) National Research University, Moscow, Russia

gadolina@mail.ru

Abstract. In this paper, the authors present some considerations on how to develop studies of wear (wear resistance) of mining machines for open-pit mining. First, the need to continue research aimed at adapting the A.K. Reisch's approach was noted to clarify and improve the empirical formula to assess the wear rate of elements of mining and road machines. Secondly, it is planned to deepen research to assess the correlations between the properties of materials used in mining engineering and characteristics of their wear resistance. Thirdly, the problem of wear of elements of mining machines is considered from the standpoint of managing the phenomenon of wear. For this, the A.K. Reisch's empirical formula can be used, and the optimization problem formulated in a general form, which is reduced to maximizing the productivity of the machine with restrictions on the rate of wear of its elements and the operating costs. Fourth, the relationship between the problem of managing the phenomenon of wear and the dynamics of mining machines is noted. Some of these researches are illustrated by examples of a qualitative nature.

1. Introduction

As long as humanity does not master thermonuclear energy sources, it is doomed to use traditional minerals. Minerals are extracted from the ground in both open-pit and underground mining operations. It is obvious that now all the main stages of mining are mechanized. Machines, which involved in mining, are characterized by complexity, size, weight and high cost. The operation of these machines often takes place in the most difficult climatic and mining conditions. Management of these machines is most often carried out by a person, with all the resulting pros and cons.

One of the main problems that stand in the way of effective use of the entire class of mining machines is the presence of friction forces, which, in particular, cause the phenomenon of wear of elements and connecting parts of mining machines.

Mining operations and equipment for them have been known to people for more than one century. And all this time, people are trying to reduce the effect of friction and wear. Basically, it has always been dedicated to finding new design solutions for machines, technological methods of their application, and the use of new materials. In the last 50 - 60 years, theoretical scientific research has also appeared, often with a capital experimental base. These studies are aimed at establishing the features of the process of friction and wear in the machines themselves, followed by generalization of private research in the structure of the General theory. Unfortunately, it was not possible to create a unified General theory of friction and wear of elements of mining machines. At the same time, due to the well-known events of the 90s of the last century, most of the experimental materials were lost in our country.

Known from the numerous literature approaches to the assessment of friction and wear in mining engineering mostly are reduced to the technique for underground mining. There are significantly fewer studies of open-pit machines. There are no experimental or theoretically based formulas in the published works.

Also, we are not aware of the work aimed at controlling the processes of friction and wear of mining machines. Let's try to talk about this on the example of mining excavators and the recommended for them in our works (e.g. in [1]) empirical formula of Reish A. K. [2] for estimation of wear rate of teeth and the cutting edge of the bucket teeth of rotary wheel.

2. Formula for wear rate estimation

Now we recall the General form of the Reish A. K. formula [1, 2]:

$$\gamma = (A \cdot P \cdot K_{p0} \cdot K_{vp0} \cdot f \cdot s \cdot \cdot t_{\mathrm{P}} \cdot K_{\mathrm{U}} \cdot K_{\mathrm{ABR}} \cdot \frac{1}{K_{\mathrm{IZN}}}) \cdot K_{t20} \quad (1)$$

where A is the proportionality coefficient; P is the pressure on the working surface of the tooth; K_{p0} is the coefficient taking into account the influence of pressure changes; K_{vp0} is the coefficient taking into account the influence of frequency of pressure change; f is the friction coefficient; s is the path of friction of the tooth; t_P – the duration of the digging; K_U is the coefficient taking into account the blunting of the tooth; K_{ABR} is the coefficient of the abrasiveness of the soil; K_{IZN} is the coefficient of wear resistance; K_{t20} is the coefficient taking into account ambient temperature.

This formula is purely empirical. Its author proposed it in the last century for a wide class of construction and road vehicles, based on his research over a period of more than thirty years (mainly in VNIISDM institution). However, the values of many coefficients of this formula are determined by him inaccurately enough, and when describing the coefficient A, an error was made in determining its dimension. It was also not explained how this coefficient can be determined. For this reason, the formula (1) in the form presented in the monograph [2] cannot be used in the practice of designing and operating any machines. It needs to be carefully refined to transferring this formula to other types and types of machines, for example, to machines for open-pit mining excavators.

The form of formula (1) shows that it is linear in the sense that changing the initial data will change the result proportionally. However, in reality, mining machines are essentially nonlinear systems.

There is a certain kind of dialectical contradiction – a simple linear formula with which the engineer can work and a real expressed nonlinear system. Within the framework of engineering methodology, you can close your eyes to this, but it should be noted that some of the empirical coefficients in formula (1) change non-linearly. Therefore, their definition is problematic. The coefficients in formula (1) can be divided into a number of groups based on certain criteria.

Thus, we can distinguish linear and nonlinear terms. This is useful when setting up theoretical research in research institutions. In engineering practice, it is more useful to divide these coefficients into groups based on the ease of determining them and their contribution to the assessment of the wear rate.

For example, it might look like this-the coefficients s, t_P depend on the operation of a particular machine (its operator) and on what technological scheme the unit participates in.

The coefficients K_{t20}, f, K_U, K_{ABR} and K_{IZN} depend mainly on external conditions.

The coefficients P, K_{p0}, K_{vp0} depend on how the process of interaction of the working body with the wear medium proceeds.

The last group of coefficients are "Terra Incognita" for modern science. Some works in this area are contradictory and not very informative [3-5, etc.]. You have to choose which of the previous authors can be trusted and rely on their ideas. In fact, this is a large and important area for future fundamental research.

A special place is occupied by the proportionality coefficient A. It is precisely its precise selection that can allow us to control the wear rate at the first stage of our research. However, its determination is still possible only for specific mining and geological and temperature conditions when analyzing the operation of a specific machine. In the process of accumulating our knowledge about the wear process of specific types of mining machines in typical operating conditions, we should strive to exclude this coefficient from formula (1).

It is important to note that a number of empirical coefficients significantly depend on the quality and properties of materials, as well as metallurgical technologies for their manufacture. Thus, different properties of steel, that have some acceptable variation in the manufacture of a machine at the factory, can greatly affect the performance of the same machine, made at different times, from different batches of steel, but for the same working project.

This requires setting and considering the problem of analyzing the influence of the spread of material properties. In addition, it is of great importance to study the influence of impurities added to steel by searching for linear regression coefficients.

3. Analysis of factors interaction

As an example, we note that it is important to obtain empirical dependencies for the refinement of the wear rate constants. In [6, 7], it is stated that "hardness will be an integral component in the criterion for evaluating the wear resistance of steel during mechanical wear". An ambiguous effect of strength was also noted. Reliable confirmation of this information is complicated by the problem of setting up a laboratory experiment with varying factors, since they are often mutually dependent. Based on the materials of [6, 7, etc.] figure 1 shows some of the graphical pair correlations of the factors.

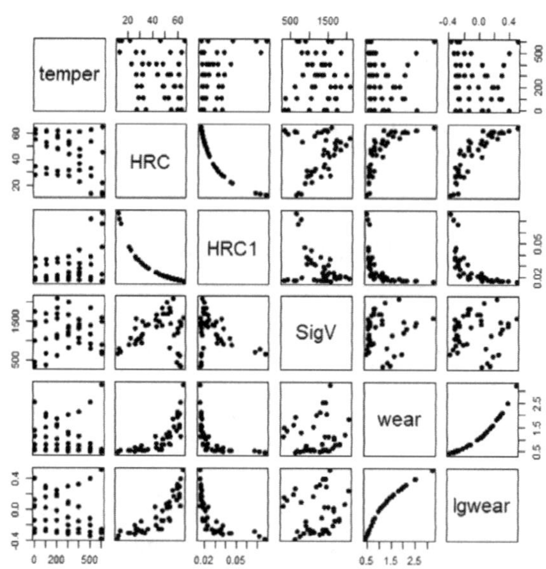

Figure 1. Pair correlations of the factors.

The graph shown in figure 1 shows paired correlations of factors for a sample of 55 elements borrowed from table in [6, 7]. Notation: temper - temperature of steels, ^{0}C; *HRC* - hardness, *HRC1*- inverse value of hardness, HRC^{-1}; *SigV* - tensile strength of steel samples, MPa, *wear*-wear rate; *lgwear* - logarithm of the wear rate. The graph is built in the R programming environment [8].

The transformations for the values of hardness and wear rate were an attempt to linearize the bonds. However, there are no obvious correlations for any pair. The tempering temperature factor was excluded from further analysis.

Next, a linear multiple regression analysis was performed, in which the wear rate was used as the response. Based on the data from [6, 7, etc.], abbreviated due to the lack of exhaustive information in some experiments, equation (2) was obtained:

$$Wear = 0.37 + 0.044 HRC - 0.00085\, SigV \tag{2}$$

The analysis, without claiming to be deeply scientific, demonstrates the approximate nature of the dependence of the wear rate on the most significant factors. The obtained dependence can later be used in planning an experiment to analyze the influence of technological factors on the quality of products according to the wear rate criterion.

An example of successful application of experiment planning using the Taguchi method in the problem of optimizing the mechanical and tribological properties of composites is given in [9]. The authors managed to optimize the homogeneity of the mixture in a 3D printer by specifically varying three parameters: the speed of the extruder screw, the temperature of the layer application, and the layer volume.

4. Optimization

The controlling the wear rate of mining machine elements is closely related to their performance (P). Thus, by increasing the speed of movement of the working body, we usually increase the wear rate, but we win in productivity. However, operating costs increase (ΔZ). Obviously, there is a need to find some compromise.

Now we write the optimization problem in the form, which is possible at present:

$$\begin{aligned} &P \to MAX \\ &MIN \le \gamma \le MAX \\ &\Delta Z \le MAX \end{aligned} \tag{3}$$

Here we assume that the performance of the excavator should be maximum with restrictions on extreme wear limits (since the range of speed control of the rotary wheel is limited) and on maximum operating costs.

Note that at present it may be necessary to provide not the maximum performance of the machine, but the desired one within its possible extreme limits. However, this does not prevent us from using the formulation (3), since, in fact, only the value of the optimization parameter MAX changes.

Note that statement (3) makes sense for organizations that operate the machine; in the design office, there problem can be written more accurately and at a higher scientific level – for example, a complex and nonlinear function of operating costs requires a separate in-depth study. The same applies to the drive of an individual car: if in the design office it can be changed in a wide range, in practice this is impossible.

Due to this reason, the problem of wear rate management is closely connected with the analysis and optimization of the drive of machines. So, for the excavator – dragline winch moving equation was obtained:

$$J\ddot{\varphi} = M_0 - \frac{\lambda}{\omega - \dot{\varphi}} - R \cdot (P_0 + R \cdot \varphi \cdot (P_0 + E \cdot F)) - R \tag{4}$$

Here: J is the moment of inertia brought to the shaft of the winch drum;

$\ddot{\varphi}$ is the second time derivative of the angular movement of the drum;

R is the diameter of the drum on the rope winding layer;

P_0 is the minimum total force in the traction ropes, at the beginning of its movement;

E is elastic modulus of the rope;

F is the cross-sectional area of the rope (total for two branches);

φ is the value of the angular movement of the drum;

$\dot{\varphi}$ is the first time derivative of the angular movement of the drum;

R is a complex function that describes the behavior of the value of the digging resistance force on the bucket teeth;

M_0, ω and λ are some parameters of the mechanical characteristic.

This equation shows the nature of a complex relationship between the forces of soil resistance to digging applied to the bucket teeth and the parameters of the traction winch drive. By changing the latter, we can influence the wear of the teeth, since it is closely related to the efforts of soil resistance to digging.

5. Conclusions

This paper identifies some possible and important areas for further research on mining machine wear, namely:

- for the first time in our country, the problem of friction and wear control has been formulated for mining machines;
- coefficients are specified in the formula of A. K. Reish, which can serve as a basis for managing wear in the practice of operating mining machines;
- the task of optimizing the performance of a mining machine with restrictions on the permissible operating costs and the wear rate of elements of mining machines is set;
- given the close relationship between the wear and the dynamics of the actuator;
- based on reference data, the influence of operational factors on the wear rate is analyzed;
- the problem of accounting for and influence of stochastic properties of materials used in mining engineering and the impact of this on wear is outlined.

References

[1] Gadolina I V, Pobegailo P A, Kritsky D Yu and Papich L 2019 Refinement of the engineering methodology for assessing the wear rate of elements of working bodies of excavators *Reliability* **68** pp 18-23

[2] Reish A K 1986 *Increasing the wear resistance of construction and road machines* (Moscow: Mashinostroenie) p 184

[3] Chudnovsky V Yu 2002 *Mechanics of Rotary Eexcavators* (Jerusalem: MIKA K. A. publishing House) p 329

[4] Novikov A 2019 *Ensuring the Durability of Working Bodies of Tillage Machines* (Moscow: INFRA) p 155

[5] Vetrov Yu 1971 *Cutting of Soil by Earthmoving Machines* (Moscow: Mashinostroenie) p 360

[6] Vinogradov V N, Sorokin G M and Kolokolnikov M G 1990 *Abrasive Wear* (Moscow: Mashinostroenie) p 224

[7] Vinogradov V N, Sorokin G M and Albagachiev A Yu 1982 *Wear on Impact* (Moscow: Mashinostroenie) p 193

[8] R Core Team 2020 R: A language and environment for statistical computing. R Foundation for Statistical Computing, Vienna, Austria

[9] Dontsov Yu V, Panin S V, Buslovich D G and Berto F 2020 Taguchi optimization of parameters for feedstock fabrication and FDM manufacturing of wear - resistant UHMWPE - based composites *Materials* **13**

Evaluation of the influence of lubricant quality on dynamic performance and resource of involute gears

S A Polyakov[1,2], **E M Kuleshova**[1], **L I Kuksenova**[2,1] **and A V Medovshchikov**[1]

[1] Bauman Moscow State Technical University, Moscow, 105005 Russia
[2] Institute of Machines Science, named after A A Blagonravov, Russian Academy of Sciences, Moscow, 101990 Russia

serpol50z@rambler.ru, lkukc@mail.ru, kuleshova.em@mail.ru, alexmed@bmstu.ru

Abstract. The article deals with the relationship between the occurrence of self-oscillations and the wear process in involute gears. The dependence of the amplitude of the torque fluctuations on the high-speed shaft of the involute gears on the type of lubricant is presented. The calculation of the transmission service life is given, taking into account the rate of growth of the dynamic coefficient.

1. Introduction

The service life limitation of involute gears can occur due to wear, due to the fact that wear leads to a change in the shape of the teeth, an increase in gaps and a significant deterioration in the dynamic performance of the transmission [1]. Therefore, the consideration of involute gears for wear resistance is an urgent task.

Dynamic parameters of involute gears will be defined in this article to assess the causes of wear and its permissible limits when using different compositions of lubricants.

2. Methodology

The service life of an involute gear according to the wear criterion can be determined using the formula [2]

$$T = \frac{h_{\lim}}{\left(\Delta h_1 + \dfrac{\Delta h_2}{U}\right) v},$$

where $h_{\lim}$ — maximum permissible wear, U — gear ratio, Δh_1, Δh_2 — values of gear and wheel wear.

The key value among the listed values is the maximum permissible wear $h_{\lim}$, for which there are no methodological grounds for choosing a specific value for involute gears. This value is limited by the growth of the dynamic coefficient K_d, which is traditionally used to evaluate the dynamism of drives. The coefficient of dynamism depends, all other things being equal, on the gap in the engagement, which is associated with the wear process almost linearly. In this paper, we will show how the introduction of a film-forming additive to a lubricant affects the change in the dependence of the dynamic coefficient on the load.

The dynamic coefficient can be determined in accordance with [3] using the formula

$$K_d = 1 + \Delta T / T_n = 1 + \left(C_f / T_n\right) \sum A_i , \tag{1}$$

where T_n is the nominal torque, ΔT is the excess of the nominal torque, C_f is the coefficient of torsional stiffness of the transmission, and A_i is the amplitude of torsional vibrations of the transmission. The traditional understanding of this value is that the increment of the torque ΔT occurs as a result of changes in the external load. However, such an overload should generally be considered as a random short-term deviation, and the nominal torque should be selected as the maximum permissible long-term current.

The most interesting is the behavior of the torque on the high-speed shaft, which plays the role of the friction force in relation to the braking torque. The dynamics of the torque on a high-speed shaft will depend significantly on the friction force in contact, thus on the internal dynamics of the transmission.

The dynamic coefficient for a high-speed shaft is proposed in [4]. It can be entered as the dynamic coefficient of a high-speed shaft, represented as

$$K_d = 1 + \Delta T_{em} , \tag{2}$$

where ΔT_{em} is the absolute value of the increase in the tad moment during vibrations, which, as shown in [4], is formed mainly due to self — oscillations, correlated with the value of T_{tor}, which is the current value of the braking (loading) moment at the reached loading stage. Formula (2) is a generalization of (1) for the case of a changing value of the braking torque and shows how the amplitude of vibrations can change due to an increase in the braking torque, including in the case of exceeding the nominal torque. The proposal of this formula is related to the phenomenon of self-oscillations that occur and may increase due to an increase in the friction force in the transmission contact due to an increase in the braking torque. As was shown in the study of self-oscillations, it is convenient to use the fast Fourier transform method to analyze the frequency response function (FRF). The use of the relative amplitude (ΔT_{em}, %), that determines the dynamic coefficient of the high-speed shaft in accordance with the formula (1) not only evaluates the dynamism of the transmission with different lubricants, but also ranks these materials by their extreme pressure action, depending on the value of the loading moment, with a given maximum dynamic coefficient.

3. Results

Thus, we obtained distributions of the amplitude of the high-speed shaft moment vibrations over the vibration frequencies when using different oils (figures (1-4)).

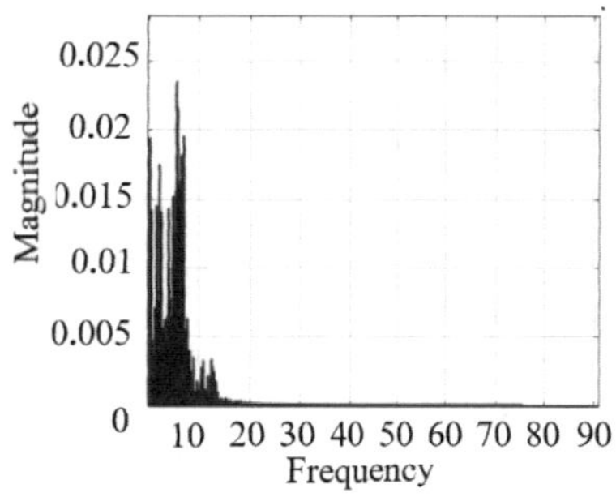

Figure 1. Dependence of the amplitude of the moment fluctuations of the high-speed shaft of the involute gear on the frequency when using mineral oil ($T_{tor} = 2$ N·m).

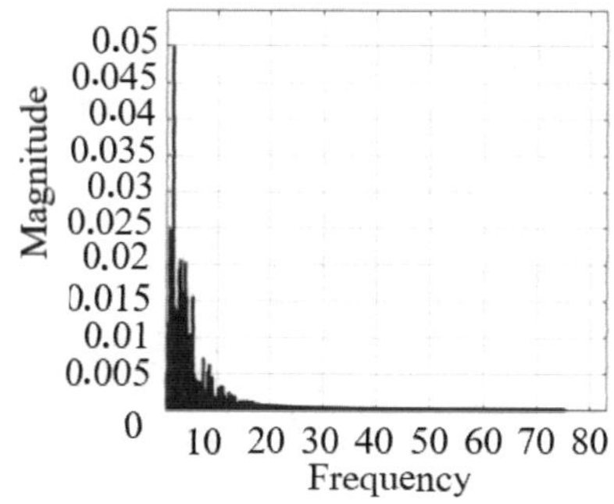

Figure 2. Dependence of the amplitude of the moment fluctuations of the high-speed shaft of the involute gear on the frequency when using mineral oil ($T_{tor} = 4$ N·m).

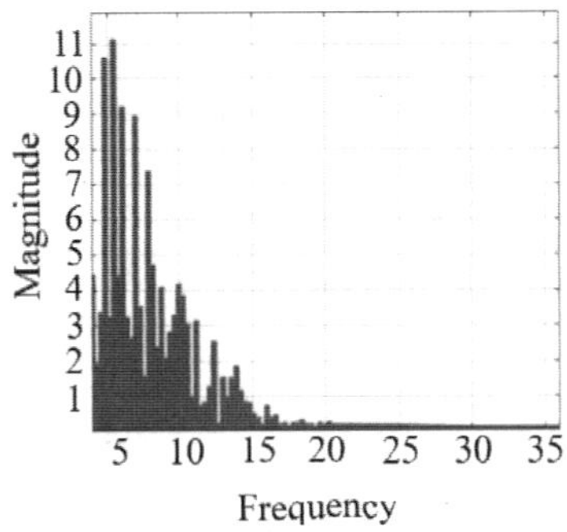

Figure 3. Dependence of the amplitude of the moment fluctuations of the high-speed shaft of the involute gear on the frequency when using mineral oil with the Striboil alloying element (T_{tor} = 2 N·m).

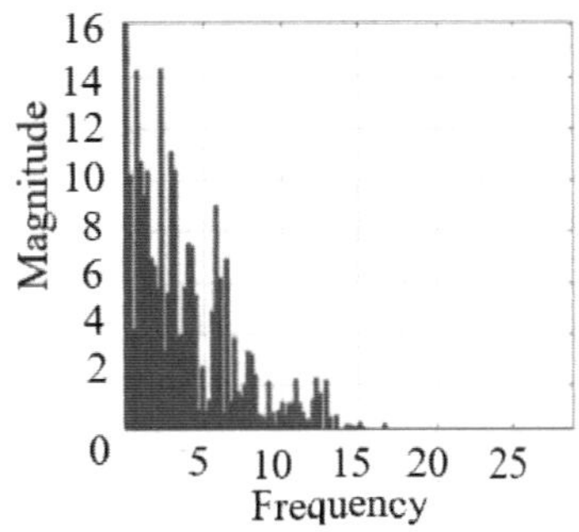

Figure 4. Dependence of the amplitude of the moment fluctuations of the high-speed shaft of the involute gear on the frequency when using mineral oil with the Striboil alloying element (T_{tor} = 4 N·m).

Analysis of the dependence of the amplitude of the moment fluctuations on the high-speed shaft on the braking moment allowed us to construct the graph shown in figure 5.

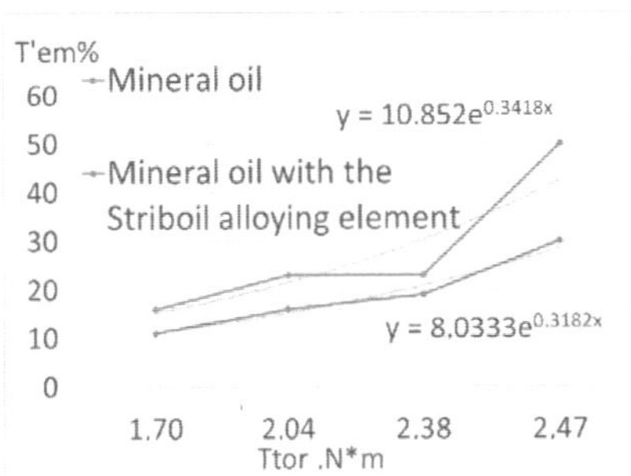

Figure 5. Analysis of the dependence of the amplitude of the moment fluctuations on the high-speed shaft on the braking moment. $T'_{em} = \frac{T_{max}}{T_{em}} * 100\%$, where T_{max} corresponds to the maximum value of the magnitude of the moment oscillation.

The graph shows, firstly, that the growth of the brake torque on the fast shaft, expressed in relative units, is growing. As the load increases, the influence of tangential slippage and sliding friction in the gearing grows, hence this phenomenon is consistent with the Hercy–Striebek diagram due to the fact that the T_{em} corresponds to the friction force [5] (as the load increases, the sliding friction coefficient grows exponentially under boundary lubrication conditions). Secondly, as the antifriction properties of the lubricant increase, the T_{em} amplitude decreases, so, in accordance with the formula (2), the dynamic coefficient of the high-speed shaft decreases, which, in turn, leads to an increase in the maximum permissible wear and an increase in the service life according to the wear criterion.

The data presented in figure 5 allowed us to construct analytical dependences of the dynamic coefficient on the load (figure 6).

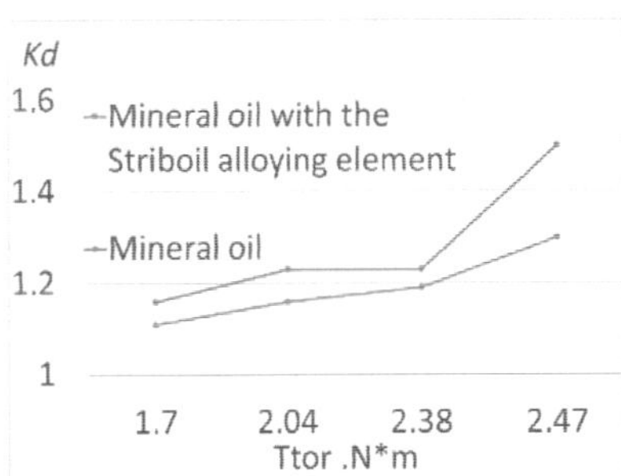

Figure 6. Dependence of the dynamic coefficient of the high-speed shaft on the braking torque.

We can calculate the service life of an involute gear using the dependencies shown in figure 6, taking into account the growth of the dynamic coefficient. Due to the fact that the value of the maximum allowable wear varies inversely with the growth of K_d, we present the proposed dependence

$$T = \frac{h_{\lim}}{\left(\Delta h_1 + \frac{\Delta h_2}{U}\right) v K_d}$$

Here are examples of calculating the service life of an involute gear with a braking torque of 2.47 N·m according to the standard method and the method that takes into account the growth of K_d when using clear mineral oil and mineral oil with the Striboil alloying element. Typical values of the parameters of an involute gear: number of teeth z_1=20; z_2=66; modulus m=1.5; $h_{\lim}=0.6\cdot m=0.6\cdot 1.5=0.9$ mm. Wear rate when using mineral oil $J=10^{-8}$, when using additives $J=10^{-10}$. The calculation results are shown in table 1.

Table 1. Service life value of an involute gear.

Calculation formula	T, hour (mineral oil)	T, hour (mineral oil with the Striboil alloying element)
$T = \frac{h_{\lim}}{\left(\Delta h_1 + \frac{\Delta h_2}{U}\right) v}$	3453	345300
$T = \frac{h_{\lim}}{\left(\Delta h_1 + \frac{\Delta h_2}{U}\right) v K}$	302	265615

Calculations show that the actual value of the service life is slightly lower using the method that takes into account changes in the rate of growth of the dynamic coefficient than using the standard method. The service life is low without the use of a film-forming additive, and it may not meet regulatory requirements, due to the fact that for such transmissions, the service life is usually at least 10 thousand hours. The service life is significantly increased due to the introduction of additives. This phenomenon is a result of an increase in K_d, which leads to a proportional increase in the actual load and a subsequent increase in the wear rate.

4. Conclusions

The dependence of the amplitude of the torque fluctuations on the high-speed shaft of an involute gear on the type of lubricant is revealed. It is shown that as the antifriction properties of the lubricant increase, the magnitude of the amplitude decreases, that leads to a decrease in K_d and, consequently, to an increase in the service life according to the wear criterion, this is confirmed by calculations.

References

[1] Pavlov V G and Ryakhovskij O A 2003 Calculation estimation of service life of evolvent straight tooth gear based on wear criterion *J. Frict. Wear* **24(3)** pp 235-41

[2] Pavlov V G 2010 Effect of contact-kinematical slippage in toothing on deterioration and efficiency of cylindrical involute gears *J. Mach. Manuf. Reliab.* **39(4)** pp 371-4

[3] Andrienko L I, Baikov B A, Zakharov M N, Polyakov S A, Ryakhovskij O A, Tibanov V P and Fomin M V 2014 *Machine Parts* ed Ryakhovskij O A (Moscow: Moscow State Technical University Press) p 472

[4] Polyakov S A, Goncharov S Y and Kuksenova L I 2016 Development of method for evaluating performance of worm gears via analysis of their tribological characteristics *J. Mach. Manuf.*

Reliab. **45(5)** pp 458-63
[5] Kalinin D V and Temis Y M 2018 An analysis of the influence of friction forces on dynamic loads in gears *BMSTU J. of Mechanical Engineering* **6(699)** pp 32-43

IOP Conf. Series: Materials Science and Engineering **996** (2020) 012021 doi:10.1088/1757-899X/996/1/012021

Computer model of contact of rough surfaces of friction material in the friction nodes of lifting and transport machines

A M Romasko and X T Do

Bauman Moscow State Technical University, Moscow, 105005 Russia

Sweetdream8190@gmail.com

Abstract. In this paper, there is proposed computer model of the contact of friction materials in the brake devices of building road machines and lifting transport machines. The contact surface is modelled by a microrelief composed of a set of elements in the form of rods. As a result of modelling, there are such contact characteristics as rapprochement, distribution of contact pressure, real contact area and the gap between the surfaces. The obtained dependencies are compared with the finite element method (FEM).

1. Introduction

When developing friction units of hoisting-transport machines and selecting friction materials for them, physical modeling of working conditions of materials on laboratory stands or friction machines is widely used. The main problem of physical modeling is the substantiation of the identity of the loading of friction materials in laboratory tests and natural conditions. This rationale is based on the correspondence of the external parameters of the friction process, most often such as: sliding speed, pressure on the nominal contact surface. In this case, the pressures on the real contact area in full-scale and on model specimens can significantly differ, with the same external parameters. Under other equal conditions, the more the sizes of the friction surfaces differ, the greater likelihood of a mismatch between the results of the friction process on small samples and the results of friction materials in full-scale friction nodes there is. It presents that the comparison of laboratory and field tests can be more adequate if when modeling the pressure distribution over the real contact areas is compared, rather than the average pressure over the nominal areas of the friction surface. This is possible, if deformations of microroughness under load are known.

Contact interaction of surfaces is the subject of contact mechanics [1-3]. When calculating the parameters of contact interaction, different surface models are used such as analytical, numerical. Examples of analytical models are the model of N B Demkin [1] and the model of J Greenwood and J Williamson. In this model, the surface roughness is modeled by hemispheres with the same radii of curvature. The distribution of irregularities in height relative to the plane of contact follows the Gaussian distribution function. Numerical models [4] in comparison with analytical ones allow taking into account additional factors and clarifying the description of contact interaction processes. This refinement is related to the distribution of sizes and shapes of individual irregularities of the contacting surfaces, the mechanism of deformation of the material of the surface layer, etc. Analyzing the existing models, it should be noted that they are difficult to apply for engineering calculations, since either they do not provide the necessary accuracy of the contact area calculations, or they have a complex mathematical apparatus.

This article discusses computer models based on technical surfaces in C++ that differ from the ideal geometric shape in the form of roughness. The model is intended, first of all, to calculate the load on the contact, which is to calculate such mechanical characteristics of the contact as the magnitude of the contact deformation, the real contact area, the pressure distribution on the contact surface.

2. Model description

The model of the microsurface of the contact is represented by a set of elements in the form of rods located next to each other. The nature of the deformation of the protrusion changes during loading, the protrusions are first deformed elastically, then elastoplastically and, finally, plastically. The real contact points are crushing areas of individual protrusions. When deriving the calculated ratios, the following preliminary assumptions are made.

The rough surface model element is located on a rigid base. If a system of concentrated forces with a step s is applied to the surface of the body, then according to the Saint-Venant's principal at a distance from the surface, commensurate with the value of s, the stress distribution is the same as if a load distribution acted uniformly on the surface. Consequently, the dimensions of the layer with an inhomogeneous strain distribution are of the same order as s. It could be considered that a concentrated force acts on each contact spot, then the dimensions of this layer can be estimated as about 100 μm, which is more than an order of magnitude higher than the roughness height. The stress in it is much lower than those that arise on the irregularities themselves, amounting to approximately 1/100 - 1/1000 part of them. Thus, the displacement due to deformation of the underlying material layer is less than the contact displacements 10 -10^2 times than can be neglected in this calculation [4].

It is considered that close contacts do not affect each other. In practice, when using friction polymer materials, the pressure does not exceed ten kilograms per cm^2, and even more rarely, by several tens of kilograms per cm^2. In most drum and belt brakes, in which brake linings are made of friction polymer materials on a rubber, rubber resin and resin binder, pressures not exceeding 1 MPa [5] are used. Note, that at such pressures the density of contact spots is low. Thus, this assumption does not introduce significant errors in the calculations.

The contact characteristics are determined only by the shape of the gap between interacting bodies, instead of the contact of two rough surfaces, the contact of a rigid flat smooth surface with an elastic body having an equivalent roughness corresponding to the gap between unreformed surfaces can be consideredwhich greatly simplifies the task. Thus, the general case of the process of interaction of solids with smooth and equivalent rough surfaces can also be taken into account. Many works have shown that the theoretical considerations underlying this approach are generally valid [6]. One of the main results of the calculations of G. Greenwood and J. Williamson is the conclusion, which is confirmed by experiment, that for many types of rough surfaces the average radius of the contact spot (excluding surface temperature) remains constant with increasing load, while the number of contact spots and the area of actual contact increases in proportion to the load [7]. We suggest using the contact model in the form of separate rods with a constant cross-section, mainly to assess the load on the contact, therefore, the assumptions made are justified.

In the general case, the maximum contact pressure q, at which plastic deformation begins, can be represented by the expression [1]:

$$q = C\sigma_y \quad (1)$$

where C - the coefficient, depending on the shape of the surface; σ_y - the yield strength of the material.

According to [2], it has a value in the range from 1.075 to 3. In the N B Demkin's works, the value of C is close to 3. When calculating the actual pressure, we can take equal microhardness $q = H$ is used, where H is the maximum Meyer's hardness [8].

To create the model, the following tasks need to be done: determination of static characteristics describing microtopography of the surface; determination of the dependence of relative deformation on pressure; finding the deformation of the totality of irregularities.

2.1. Microsurface model

When generating a microsurface, the input parameters are the standard characteristics of microtopography of the contacting surfaces by GOST 2789 – 73.

The entire surface is divided into separate areas with a base length l. In numerical simulations, this separate area is approximated by an n x n matrix of bar elements having the same rectangular section, but different heights.

When generating a microsurface, the heights of the rods at the base length are satisfied with the following conditions:

Condition 1: The value is accepted in the range from 0 to R_{max};

Condition 2: The arithmetic mean of the differences in the heights of the rods and (R_{max}-R_p) in each row of the matrix must be equal to R_a;

Condition 3: The average value of the pitch of the profile irregularities within the base length is S_m;

Condition 4: The relative value of the sum of the lengths of elements whose height is greater than or equal to (R_{max} - R_p) to the base length l is t_m;

Condition 5: The relative value of the sum of the lengths of elements whose height is greater than or equal to $0.8R_{max}$ to the base length l is equal to t_{20}.

A microtopography model of a surface that satisfies all these conditions reproduces the real surface well, but the formation of the model takes a lot of time. Therefore, to save time, you can apply one of the following sets of conditions:

K1: Condition 1, Condition 2, Condition 4, Condition 5;

K2: Condition 1, Condition 2, Condition 3;

K3: Condition 1, Condition 2, Condition 4.

2.2. Model contact

It is assumed that one of the materials of the contact pair has significantly lower mechanical characteristics compared to the other. This means that only softer material is deformed. This corresponds to friction brakes and couplings in which steel or cast iron and friction plastics contact.

Features of the model:

- it is assumed that during loading the friction material is deformed first elastically, then elastoplastically, and when the pressure in the contact reaches the maximum Brinell hardness - plastically;
- it is assumed that the contact area is determined by the mechanical characteristics of the friction material (softer);
- the mechanical characteristics of the material at different temperatures should be experimentally determined in advance;
- the contact of rough surfaces is discrete.

The main mechanical characteristics for use in the model are obtained from the experiments of the considered materials for compression by GOST 4651-2014 (ISO 604: 2002). According to the test results, the dependence function $\sigma = f(\varepsilon)$ is created using the cubic spline. The program is implemented in C ++ [9]. In the model, the contact pressure at which plastic deformation begins, is equal to HB. It denotes the relative deformation in this case ε_k.

The computer model of the contact of rough surfaces works according to the following algorithm. For a given common normal load on a friction pair, the relative deformations of each protrusion are determined by the characteristics of the surface microgeometry. If the found value is less than ε_k, then the contact stress is found by the function $\sigma = f(\varepsilon)$;if it is greater than or equal to ε_k, then the contact stress is equal to the hardness HB. For each protrusion, depending on its area and contact pressure, a load is determined. The sum of loads of all the protrusions should be equal to the specified load.

3. Result and discussion

Giving an example of modelling a rough surface by preliminary grinding, the roughness parameters according to the results in the book [1] are presented in table 1.

Table 1. Parameters of surface microgeometry.

R_{max} (µm)	R_a (µm)	R_p (µm)	S_m (µm)	l (mm)	t_{20}	t_m
10	2.5	5	160	0.8	0.1	0.5

Figure 2 shows the curves of the supporting surfaces according to the formula presented by professors N B Demkin, I V Kragelkim and according to the sets of conditions K1, K2, K3. The curve calculated in our program according to the set of conditions K2 best fits the curve calculated by the formula recommended in [1].

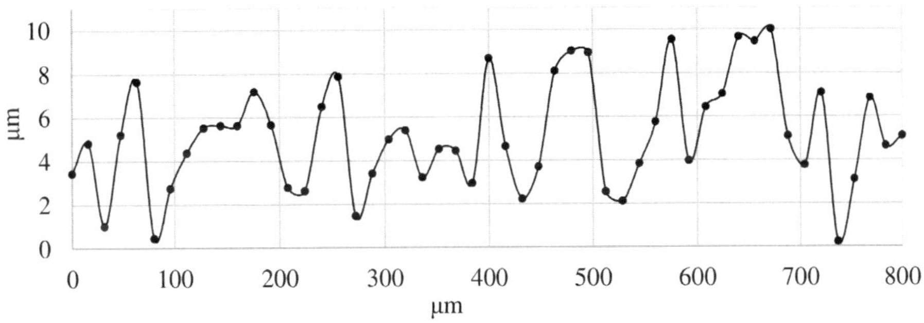

Figure. 1. Surface profilogram model.

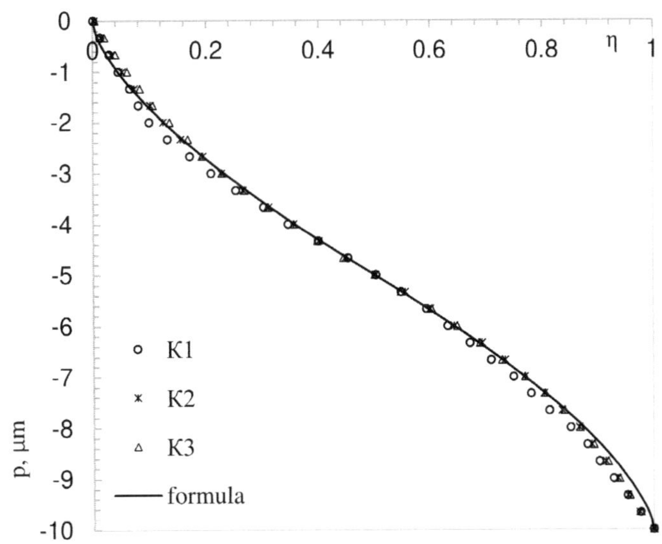

Figure 2. Abbott curve.

As examples of the contact model, the contact model of the rough surface of the friction material 2339.62 with hardness HB = 23 MPa (the brakes of trucks) with a smooth steel surface is simulated. The stress diagram of material from a compression experiment at 20°C on machine QUASAR 50 is shown in figure 3. On table 2, the stress distribution over the actual contact areas at various nominal pressures is presented: 0.1; 0.5; 2.0; 3.0 MPa.

Figure 4 presents the simulation results, namely: the influence of the nominal pressure q_a on the ratio of the real area to the nominal A_r/A_a; ratio of rapprochement to maximum roughness height p/R_{max}; real maximum pressure q_r and gap between the surfaces.

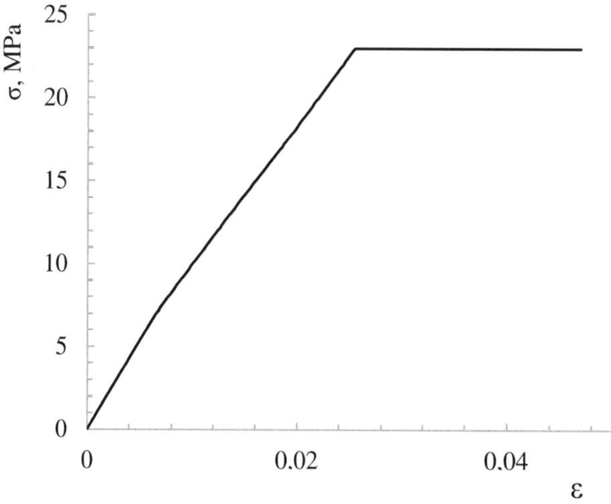

Figure 3. The stress diagram of material.

Figure 4 confirms that in the range of generally accepted values of nominal pressures for this material, the real contact area is hundredths and tenths of the nominal area, which corresponds to the results given in many works. From the moment when the nominal pressure reaches 2.4 MPa, the maximum pressure on the real contact area changes little and corresponds to the hardness of the material. This can be explained by the fact that from this moment many individual contact spots are already in a state of plastic deformation and although gradually new protrusions come into contact, the actual maximum pressure changes little. In this case, only the area of the protrusions with maximum pressure on the contact increases. As the nominal pressure increases, the rapprochement increases and the gap volume decreases.

To verify the reliability of the developed simulation model, the stress-strain state of the contact was modeled by the finite element method. Modeling was performed using the APDL program (ANSYS). In figure 5 the dependences of the ratio p/R_{max} on the nominal pressure are presented, where p is the rapprochement. The solid line shows the FEM calculation result; round dots - the result of modeling under the condition that the transition from elasticity to plastic deformation is the hardness HB of the watched material; triangular points - the result of the simulation under the condition of transition from elastic to plastic deformation at stress equal to $2\sigma_y$; line with rhombuses - the result of the simulation under the condition of transition at stress equal to $3\sigma_y$. From figure 4 it shows that the simulation result using HB as a criterion for the transition from elastic to plastic deformation gives results close to the results of FEM calculations.

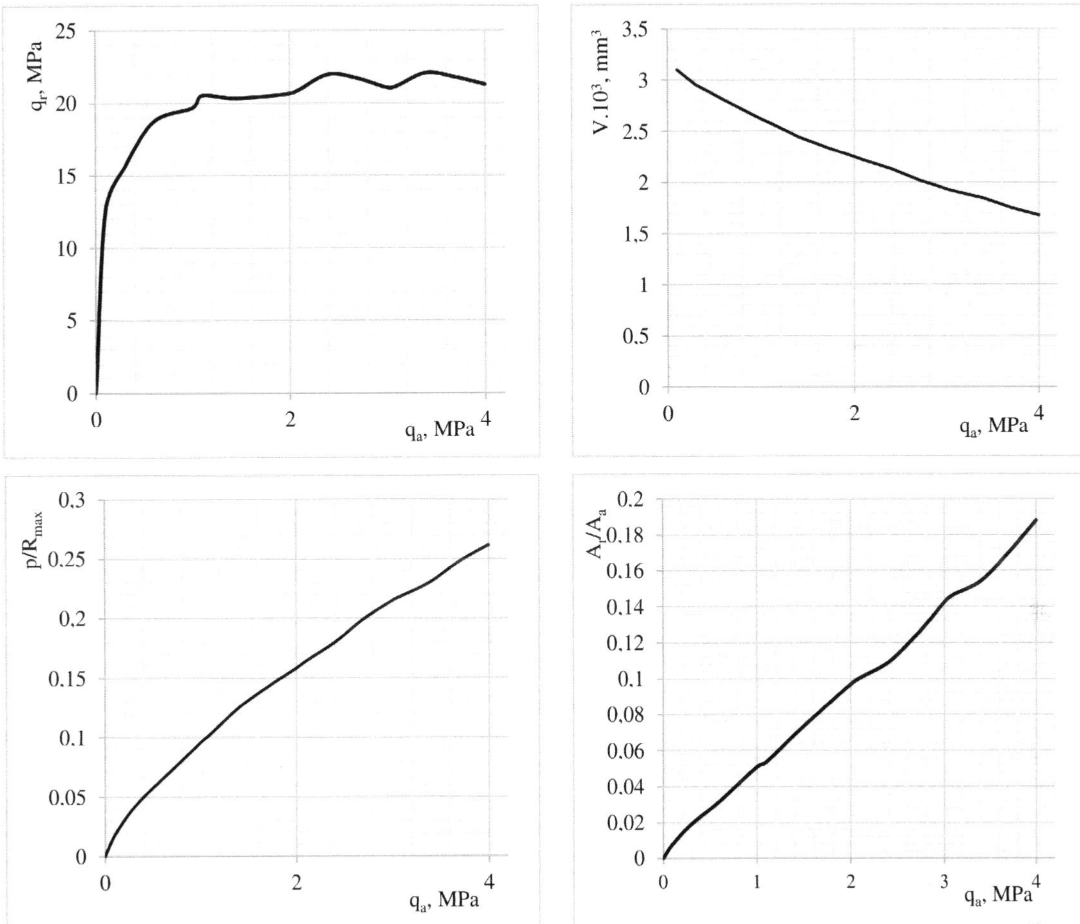

Figure 4. Effect of nominal pressure q_a on contact characteristics.

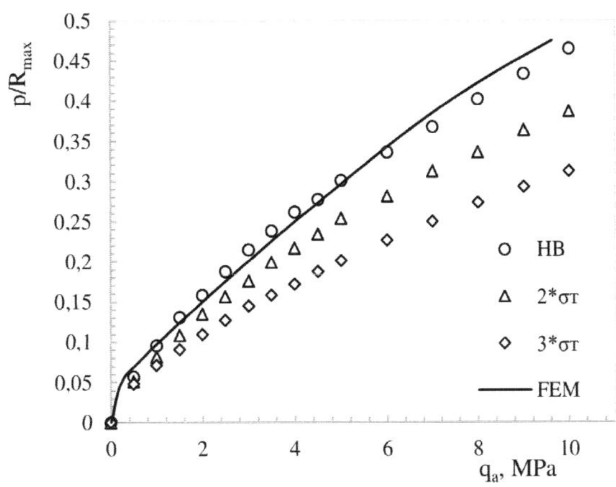

Figure 5. Comparison of simulation results with the finite element method.

IOP Conf. Series: Materials Science and Engineering **996** (2020) 012021 doi:10.1088/1757-899X/996/1/012021

4. Conclusions

Using standard characteristics of surface roughness allows to create computer models for calculating the distribution of pressure on the contact. The proposed model of the formation of the contact area allows to take into account the features of surface microgeometry and reflects the approximation of rough surfaces and the nature of the pressure distribution over microroughnesses. Such model of contact of friction material can be used to evaluate loads at actual contact areas, when collating the results of laboratory and full-scale tests of friction devices. This model of contact interaction has featured in comparison with the traditional approach. This is, firstly, the rejection of the assumption that the radii of curvature of the peaks of the bumps are the same and, secondly, the rejection of the need to know the law on the distribution of the vertices of the protrusions.In the future, the model is supposed to be used to calculate temperatures that occur during dry friction in the zone of real contact.

Notation

R_p	maximum height of the protrusion of the rough surface
R_{max}	full profile height
R_a	the arithmetic mean deviation of the rough surface profile
t_m	the relative reference length of the profile along the midline
t_p	the relative reference length of the profile, where p is the value of the profile section level, often equal to 20% of the total profile height (t_{20})
l	the base length
S_m	the average pitch of irregularities
q_a	nominal pressure
q_r	real pressure
A_a	nominal contact area
A_r	real contact area
η	relative area
ε	relative deformation
σ	contact stress
p	rapprochement
V	the gap between the surfaces
σ_y	yield strength

References

[1] Demkin N B and Ryzhov E V 1981 *Surface quality and contact of machine parts* (Moscow: Mechanical Engineering) p 224

[2] Johnson K L 1989 *Mechanics of contact interaction* (Moscow: Mir) p 510

[3] Popov V L 2013 *Mechanics of contact interaction and physics of friction. From nanotribology to the dynamics of earthquakes* (Moscow: FIZMATLIT) p 352

[4] Kragelsky I V, Dobychin M N and Kombalov V S 1977 *Fundamentals of calculations for friction and wear* (Moscow: Mechanical Engineering) p 526

[5] Hebdy M and Chichinadze A V 1992 *Handbook of tribotechnology. In 3 vol* ed Hebdy M and Chichinadze A V vol 3 (Moscow: Mechanical Engineering) p 730

[6] TikhomirovV P, Gorlenko O A and Poroshin V V 2004 *Method for modeling processes in tribotechnical systems* (Moscow: MGIU) p 292

[7] Chichinadze A V, Brown E D, Boucher N A *et al* 2001 *Fundamentals of tribology (friction, wear, lubrication): Textbook for O-75 technical universities* ed Chichinadze A V (Moscow: Mechanical Engineering) p 664

[8] Bowden F P and Tabor D 1968 *Friction and lubrication of solids* (Moscow: Mechanical Engineering) p 543

[9] Burden R L and Faires D J 2011 *Numerical Analysis* (Brooks/Cole) p 872

Metal cladding of friction surfaces in liquid media

V V Shapovalov, Yu F Migal, A L Ozyabkin, M A Burakova, V A Feizova and R A Korniyenko

Rostov State Transport University, Rostov-on-Don, Russia

valentina.feizova@yandex.ru

Abstract. On the basis of the quantum-chemical method, the possibility of using a liquid lubricant containing a fine powder of the Al-Mg-Li system is considered. It provides a high-quality process of metal cladding of steel friction surfaces and, as a result, an increase in the resource of a single application of a liquid lubricant and the resource characteristics of the tribological system "wheel rail".

1. Introduction

Strong mechanical stresses that arise during the operation of the locomotive lead to the damage to the rolling surface of the wheel-set of railway rails. The destruction of the tire, the occurrence of cracks on the rail head can lead to an emergency situation on the railway tracks. Intensive wear of rails and wheel flanges of the rolling stock is the main threat to the safe train movement.

According to the rules of technical operation of the railways of the Russian Federation, it is forbidden to include carriages on long-distance passenger trains with tire wear of more than 7 mm, local traffic of more than 8 mm, and for freight carriages of more than 9 mm. It is forbidden to use locomotives for trains with tire wear over 7 mm, and for tender tires over 9 mm [1].

Timely turning of wheel-set tires restores the working profile of locomotive wheel-sets and allows prolonging their service life [2].

The resource costs of wheels and rails can be cut almost threefold by turning wheel rims on the rolling surface. Figure 1 and 2 show the types of railroad wheel turning, where h_0' is the wheel turning depth along the rolling surface equal to no more than 10 mm, and h_0'' is the wheel turning depth along the ridge equal to at least 30 mm.

Reducing the wear of wheels and rails by 3-4 times and energy consumption for traction up to 10% is achieved by applying lubrication technologies for the wheel flanges and the side surface of the outer rail. Existing lubrication technologies are performed with the help of railroad cars, railroad locomotives, suburban trains, stationary track lubricators, traction rolling stock equipped with onboard ridge lubricators. Increasing the efficiency of lubrication is also achieved through the improvement and unification of lubricants (taking into account the regions of application) [3].

IOP Conf. Series: Materials Science and Engineering **996** (2020) 012022 doi:10.1088/1757-899X/996/1/012022

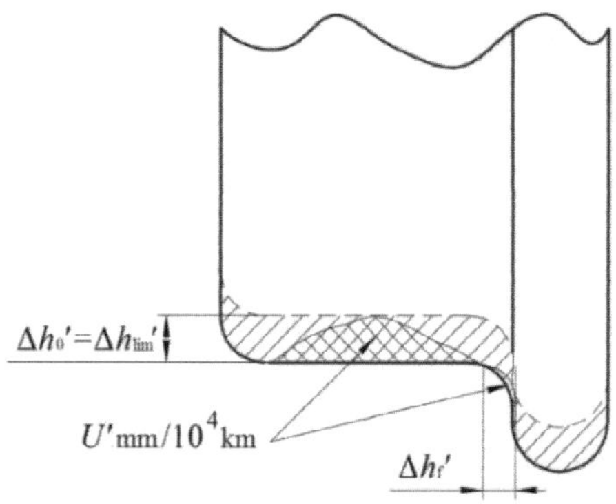

Figure 1. Turning a wheel on a rolling surface.

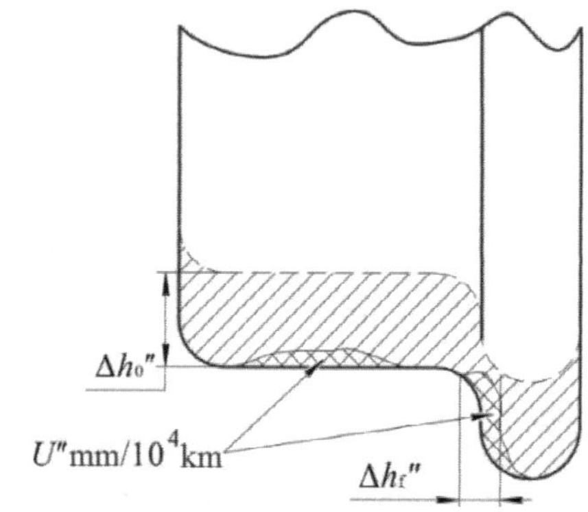

Figure 2. Turning a wheel on a flange.

The main problem with this lubrication technology is that the liquid lubricant has a short lubrication service life, i.e. it does not remain on lubricated surfaces for a long time because it flows down into the ballast prism, polluting it and everything around.

Metal-plating of steel friction surfaces in liquid media makes it possible to create modern lubricants for stationary track lubricators that increase the resource characteristics of the wheel-rail tribosystem. It means that it increases the resource of a single application of a lubricant on the friction surface [3]. Powders of soft metals and their alloys are used as cladding elements. There are processes of creating a dynamic effect on the friction system in the form of external, natural vibrations or frictional natural oscillations of contacting micro- and macro-roughness in the friction zone. In addition, there can be processes of suppression of "natural vibrations" of contacting micro- and macro-roughness. The total contact stiffness also changes when the direction of the relative sliding of the friction surfaces shifts [4].

We assume that a fine powder of the Al-Mg-Li system in a liquid medium of a lubricant will be in a suspended state, i.e. it will neither precipitate nor float due to its unique density (2.47 $g \cdot cm^{-3}$) [5].

When selecting a fine-dispersed powder based on aluminum-lithium alloys in such a way that the specific gravity of the lubricant and the aluminum-lithium powder are close in value, which will ensure high-quality metal-coating of friction surfaces in a liquid lubricant and, accordingly, will increase the efficiency of the lubrication technology with stationary track lubricators working with liquid lubricants.

On the basis of quantum-chemical methods, calculations of the strength of chemical bonds of various polyatomic objects (molecules and solids) were performed with an accuracy comparable to the accuracy of instrumental measurements of the firmness of such objects. In the case of solids, this statement refers mainly to defect-free structures that do not have local defects.

In this work, quantum-chemical methods were applied to study materials that are promising for application on the rolling surface of railway wheels. To study this issue, we carried out numerical experiments using methods based on the DFT approximation. Nowadays the accuracy of such methods is quite high, which makes it possible to consider the results obtained with their help as a fairly reliable guideline when designing various materials. In this work, we used the ADF-BAND program complex [6], which was previously used in our group to investigate the strength of steel grain boundaries [7].

2. Purpose and objectives of the study

Based on the quantum-chemical method, consider the possibility of using a liquid lubricant containing a fine powder of the Ai-Mg-Li system to ensure a high-quality process of metal plating of steel friction surfaces and, as a result, increase the resource of a single application of liquid lubricant and the resource characteristics of the wheel-rail tribological system.

3. Materials and research methods

To calculate the strength of the objects studied in this work, we used the following ratio:

$$E_{strength} = \frac{\{E(I) - E(II) - E(III)\}}{S}$$

where *E (I), E (II), E (III)* are the energies of the initial object and its two desoldered parts, S is the area of desoldering (figure 3). Thus, $E_{strength}$ is the energy required to separate a body into two parts, referred to the division surface.

It is important to note that the model used in our calculations (figure 3) is quite simple. In the direction of the vertical axis, 5-6 layers of atoms are considered, and the horizontal layers represent an infinite two-dimensional periodic structure. Figure3 displays nine cells of this structure.

To confirm the accuracy of the results obtained using this model, it is necessary to compare them with already known experimental data. We chose the following parameters for this purpose: the first one was the melting point, which integrally characterizes the strength of interatomic bonds in the metal; and the second one was the ultimate mechanical strength of metals, which also depends on the strength of interatomic bonds. The melting temperature should be considered a more reliable experimental value in this case, since it depends on the defectiveness of metals to a much lesser extent than the ultimate strength.

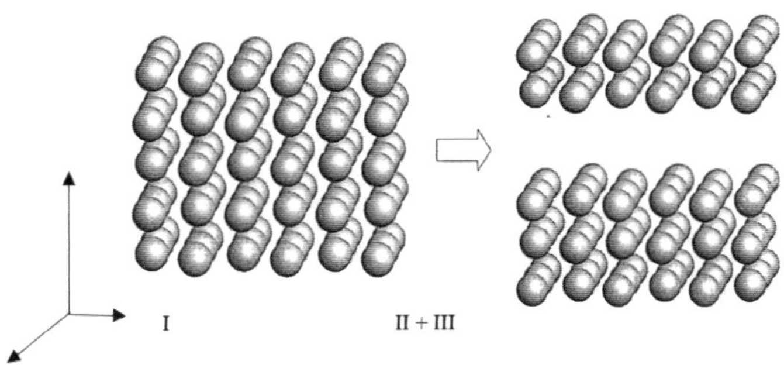

Figure 3. The disintegration of the original complex into two parts under the action of a power load.

4. Results and discussion

In each column of table 1 below you can find relative values (with respect to the maximum value of this parameter for each metal) along with experimental and calculated data.

Table 1. Calculation results and comparison with already known experimental data.

Chemical elements	Calculation - specific cohesive energy, eV/A2		Experimental data on the ultimate strength, MPa		Experimental melting point, K	
Fe	0.48	100%	250	100%	1808	100%
Cu	0.25	52%	220	88%	1356	75%
Al	0.19	40%	80	32%	934	52%
Zn	0.15	31%	-	-	692	38%

To make it crystal clear, these results are demonstrated in graphical form (figure 4). The qualitative agreement between theoretical and experimental data is obvious. It was found out that the obtained group of metals Fe, Cu, Al, Zn fully agrees with the already known experimental data, when arranged according to a decrease in their strength. This fact allows supposing that the results presented below, that have been obtained using the same model, are also close to reliable values.

5. Calculations of cohesion and adhesion in a multilayer system

As a specific example, let us consider a system consisting of several layers of atoms of various elements and their compounds. To make things clear, let us choose a system consisting of iron atoms (the base element), as well as aluminum and aluminum oxide. Our task in this case is to carry out a comparative analysis of the strength of the layers of each of the elements of the system and the intermediate layers between them. To assess the strength of the inner layers of the metal, the cohesion energy was calculated; the strength of the intermediate layers was estimated by determining the adhesion energy.

When calculating the energy of adhesion, it was assumed that the atoms of the first layers of a chemical element applied to the base metal have the same symmetry as the atoms of the base metal (space-centered cube (SCS) in the case of iron and aluminum). In addition, the diffusion of atoms of the deposited element into the surface of the main element was taken into account. For this purpose, the atoms of the applied element were introduced into the upper layer of atoms of the main element.

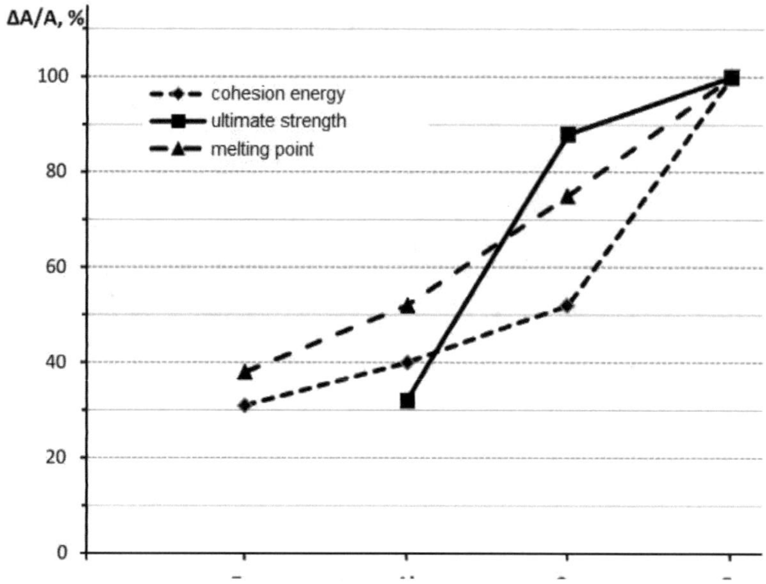

Figure 4. Comparison of the data on cohesion energy obtained using the model with experimental data on the ultimate strength and melting point of various elements.

Below, tables 2 and 3, as well as figure 5 and 6 display the results of calculations of the energy of aluminum adhesion on iron and aluminum oxide on aluminum. The rest of the results are summarized in table 4.

Table 2. Aluminum adhesion on iron (SCS symmetry).

	Lattice symmetry	Number of layers	Layer energy, eV	Energy of system (3Fe,2Al), eV	Sell dissociation energy, eV	Specific cohesive energy, eV/A2
Fe	SCS	3	-55.21	-73.77	4.19	0.36
Al	SCS	2	-14.37	-	-	-

Table 3. Adhesion of aluminum oxide to aluminum.

	Lattice symmetry	Number of layers	Layer energy, eV	Energy of system (3Fe,2Al), eV	Sell dissociation energy, eV	Specific cohesive energy, eV/A2
Fe	SCS	3	-3.05	-63.20	1.80	0.11
Al	SCS	2	-38.35	-	-	-

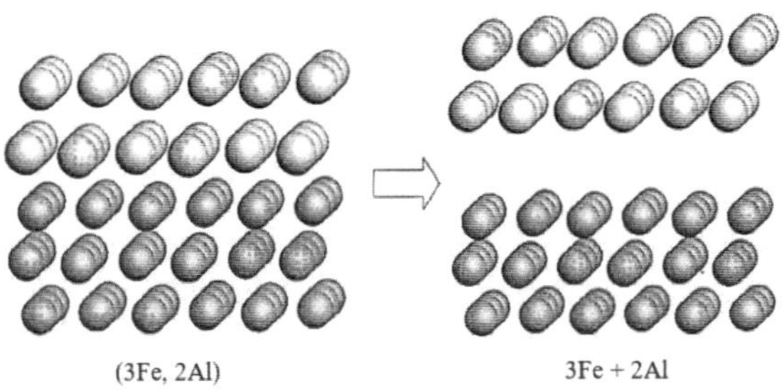

Figure 5. Decomposition of the 3Fe + 2Al complex.

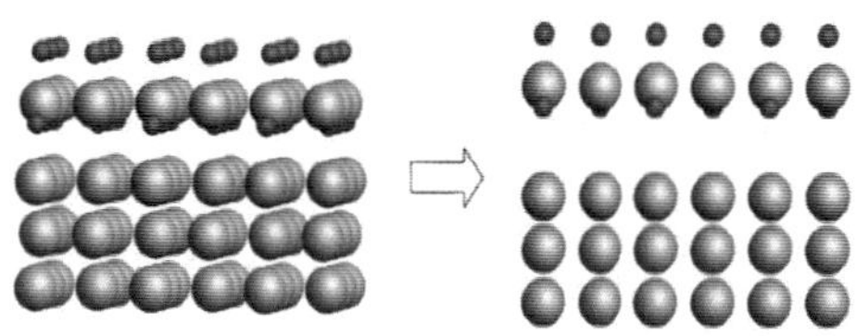

Figure 6. Decomposition of the $3Al+AlO_2$ complex.

Table 4. Summary of the obtained Results.

Elements	Bond type	Bond energy, eV/A2
Fe-Fe	Cohesion	0.48
Fe-Al	Adhesion	0.39
Al-Al	Cohesion	0.19
Al-AlO_2	Adhesion	0.11

Table 4 shows the corresponding values of energies and types of bonds for the corresponding elements of the investigated layers. As a result of the analysis of the data obtained, a consistently decreasing strength of chemical bonds in the system of layers (Fe, Fe), (Fe, Al), (Al, Al), (Al, AlO_2) is clearly visible. Thus, during friction, the outer layer of AlO_2 with the lowest adhesion energy will serve as the 'primary lubricant'. When it is wiped off, the next layers of the system will be sequentially wiped off, protecting the base element (iron) from destruction.

6. Conclusion

When using stationary lubricators on the road network of Russian Railways, it is important to use modern liquid lubricants that increase the resource of the wheel-rail tribosystem. Such lubricants include oils containing cladding elements, such as finely dispersed powders of metals and their alloys.

It has been determined that the results of calculating the strength of the studied elements Fe, Cu, Al and Zn obtained by the quantum-chemical method are fully comparable with the already known experimental data, such as cohesion energy, experimental ultimate strength and melting point.

A comparison of the calculated values of the adhesion energies of the intermediate layers and the cohesion energies of the inner layers of a multilayer system revealed a decrease in the value of the strength of chemical bonds in the considered systems of layers (Fe, Fe), (Fe, Al), (Al, Al), (Al, AlO_2).

A lubricant containing a metal-plating additive in the form of a fine powder of the Al-Mg-Li system increases the efficiency of the lubrication technology when using stationary track lubricators. The effectiveness of lubrication consists in keeping the lubricant on the rolling surfaces of wheels and rails.

References

[1] Ministry of Transport of the Russian Federation 2019 *New Rules for the technical operation of the railways of the Russian Federation* (Moscow: Morkniga) p 280

[2] JSC Russian Railways 2016 *Instructions for inspection, inspection, repair and formation of wheel pairs of locomotives and rolling stock of 1520 gauge railways* (Moscow) p 126

[3] JSC Russian Railways 2016 *Order of January 16, 2015 N 60r On Approval of the Concept of Development of the Technology of Lubrication of the Wheel--Rail Contact Zone at JSC Russian Railways*

[4] Shapovalov V 2018 Method for increasing the efficiency of the friction system Pat. of the RF № 2674899

[5] https://markmet.ru/referat_po_metallurgii/alyuminii-litievye-splavy-referat

[6] Velde G, Bickelhaupt F, van Gisbergen S, Guera C, Baerends E, Snijders J and Ziegler T 2001 Chemistry with ADF *Computational chemistry* **22** pp 931-67

[7] Migal Yu, Kolesnikov V and Kolesnikov I 2016 Impurity and Alloying Elements on Grain Surface in Iron: Periodic Dependence of Binding Energy on Atomic Number and Influence on Wear Resistance *Computational Material Science* **111** pp 503-12

Abrasive wear of metal surface modified with mineral particles

A V Skazochkin

Kaluga branch of the Russian Presidential Academy of National Economy and Public Administration, Okruzhnaya St, 4, building 3, Kaluga, Russia, 248030

avskaz@rambler.ru

Abstract. The resistance of mineral coatings created on the surface of specimens of 20X13 steel (Russian analogue of steel 20Cr13 (EU)) modified with ultradispersed mineral particles to abrasive wear during friction against fixed particles was determined, and the wear parameters of layers modified with mineral particles were compared with the parameters of surface layers created by laser cladding of alloys of different compositions. The studies were carried out on an experimental setup for testing abrasive wear under a static load of 60 N and 160 N and determined the weight wear, the average wear rate and the average wear rate of the surface layer of three samples with different mineral coatings. According to the results of measurements at a pressure on the wear surface of 0.5 MPa, a sample was determined with an indicator of weight wear comparable to a coating made of cobalt alloy, created by the method of laser cladding and about four times inferior in this indicator, to a coating of a nickel alloy, created by the method of laser cladding and tool steel. The practical use of various methods of protecting parts from abrasive wear is determined, along with the physical and technical characteristics of the layers, by the economic feasibility in specific operating conditions.

1. Introduction

The creation of protective films, coatings and modified layers with properties that differ from those of the main part is one of the promising areas of material science [1]. Evaluation of the practical benefits of the creation of films or layers traditionally consists in determining the degree of protection from the external environment (thermal cycling loads, abrasive, water, acids, alkalis, etc.) and is characterized by a set of properties and measured parameters (hardness, coefficient of friction, resistance to corrosion, wear resistance, etc.) [2-3].

Interest in modified layers enriched with microparticles of various minerals is due to their prospects as wear-resistant, corrosion-resistant, extreme pressure, antifriction coatings created on the surface of friction pairs of various metals - steels [4-6], titanium [7], aluminum [8], working in an aggressive environment - abrasive, sea water, in the presence of gases, acid solutions, under thermal cyclic loads [9, 10]. The special properties of mineral coatings, as well as such characteristics of the technology of their creation, such as the locality of deposition of coatings, low temperatures during their production, high speed of creation of coatings, give the technology additional marketing advantages in the market of wear-resistant coatings [4, 6]. Entry into the market of metallurgical, mining, construction equipment traditionally precedes research in independent laboratories and operational tests at the sites of corporate partners. The resource of equipment, machine assemblies and individual parts operating at enterprises of these industries strongly depends on abrasive wear, which occurs when parts of mechanisms are rubbed against the processed material, as well as in friction pairs operating in a fine abrasive medium [11-13].

For the equipment of enterprises in these industries, the study of the properties of coatings or protective layers created on metal parts under conditions of abrasive wear is especially relevant [14]. Until now, based on the results of the use of mineral coatings under abrasive wear conditions, created on various parts, there have been exceptionally high-quality assessments of wear resistance [4, 6]. The purpose of this article is to determine the resistance of mineral coatings created on the surface of samples of steel 20X13 (Russian analogue of steel 20Cr13 (EU)), modified with ultrafine particles of minerals, to abrasive wear when rubbing against fixed particles and to compare the wear parameters of layers modified with mineral particles with parameters of surface layers created by laser cladding of alloys of different composition.

2. Materials and methods

The essence of testing surface layers modified with mineral particles for abrasive wear when rubbing against fixed abrasive particles is that the surface of the test sample interacts for some time at a fixed load with fixed abrasive particles (abrasive bar) [3]. Before and after the end of each test cycle, the test sample is weighed and the wear resistance of the surface layer is determined.

The study was carried out on cylindrical metal samples 10 mm in diameter and 25 mm in length. On the end surface of the samples under study, layers modified with ultradispersed particles of minerals were created using the basic technology of S&PA "Geoenergetika" [4]. In total, three types of layers were created, differing in the types of mineral mixtures, under the same technological conditions, which implied a difference in tribological parameters, including the wear resistance of the samples. During the research, one sample of each type of layers was used.

The samples were studied on an experimental setup for testing abrasive wear during friction against a fixed abutment with a sample loading system consisting of a loading rod with a socket for fixing a mandrel and a rod for creating a pressure load using weight weights. The tests were carried out under the action of a static load of 60 N and 160 N, the relative error in determining the load did not exceed ± 1%.

To determine the weight wear, the test sample was weighed before and after each test cycle. When measuring weight wear, we used an electronic balance HTR-120CT with a Tuning Fork sensor, weighing range from 0.01 g to 120 g with a discreteness of not more than 0.0001 g, with a verification scale mark of at least 0.001 g and an accuracy class according to GOST 24104- 2001 special (I) (RU).

To carry out the relative wear resistance of the surface layer of the samples under study, abrasive material 12 CT 1 (RU) with an electrocorundum grinding grain with a hardness of at least R or S (hard) according to GOST 52587 (RU) and a grain size of 63-50 μm according to GOST 52381 (RU) was used. The contact surface area of the sample is 314 mm^2.

The tests were carried out by rubbing the end surface of the sample against the surface of the abrasive bar. The friction path for one test cycle was determined empirically and was constant within the entire test series. The linear velocity of the relative movement of the sample over the abrasive surface was 0.313 ms^{-1}. The choice of the magnitude of the load and the speed of movement was determined by the need to avoid heating the end surface of the samples, which affects the properties of the surface layer and the capabilities of the installation.

According to the test results, the following indicators of wear resistance were determined:

- weight wear - change in sample weight during testing, defined as the difference between the initial and final weight of the sample in each test cycle, mg;
- average wear rate - the ratio of the weight wear value to the time interval during which it arose, is defined as the ratio of weight wear to the duration of one test cycle, $mg \cdot s^{-1}$;
- average wear rate - the ratio of the wear value to the conditioned path, on which wear occurs, is determined as the ratio of the weight wear of the sample to the actual friction path traveled in one test cycle, $mg \cdot m^{-1}$.

3. Results and discussion

The results of testing the surface layers of three samples modified with different mixtures of mineral particles are presented below in tables 1-3.

Table 1. Test results of sample No. 1 for abrasive wear when rubbing against monolithic abrasive.

No. cycle	Initial weight (g)	Final weight (g)	Friction path (m)	Test time (s)	Weight wear (g)	Average wear rate ($g\ s^{-1}$)	Average wear rate ($g\ m^{-1}$)
Load 60 H							
1	29.908	29.9	9.39	30	0.008	0.00027	0.00085
2	29.9	29.898	9.39	30	0.002	0.00007	0.00021
3	29.898	29.895	9.39	30	0.003	0.00010	0.00032
Load 160 H							
4	29.895	29.874	9.39	30	0.021	0.00070	0.00224
5	29.874	29.858	9.39	30	0.016	0.00053	0.00170
6	29.858	29.844	9.39	30	0.014	0.00047	0.00149
7	29.844	29.818	9.39	30	0.026	0.00087	0.00277
8	29.818	29.795	9.39	30	0.023	0.00077	0.00245
9	29.795	29.271	9.39	30	0.054	0.00180	0.00575

By the end of cycle no. 6 the abrasive block was very "greasy" that is it lost its original roughness and was replaced with a new one. On test cycle No.8. the base metal appeared. Average values of weight wear over the thickness of sample No. 1 were 4.3 mg at a load of 60 N and 20 mg at a load of 160 N.

The test results of sample No. 2 are presented in table 2.

Table 2. Test results of sample No. 2 for abrasive wear when rubbing against monolithic abrasive.

No. cycle	Initial weight (g)	Final weight (g)	Friction path (m)	Test time (s)	Weight wear (g)	Average wear rate ($g \cdot s^{-1}$)	Average wear rate ($g \cdot m^{-1}$)
Load 60 H							
1	29.84	29.828	9.39	30	0.012	0.00040	0.00128
2	29.828	29.819	9.39	30	0.009	0.00030	0.00096
3	29.819	29.809	9.39	30	0.01	0.00033	0.00106
Load 160 H							
4	29.809	29.722	9.39	30	0.087	0.00290	0.00927
5	29.722	29.632	9.39	30	0.09	0.00300	0.00958

It should be noted that the actual wear of the surface layer enriched with minerals occurred at a load of 60 N. During the test cycle No. 4 the base metal appeared on the surface of the sample No. 2. The average weight wear at a load of 60 N was 10.3 mg.

The test results of sample No. 3 are presented in table 3.

Table 3. Test results of sample No. 3 for abrasive wear during friction against monolithic abrasive.

No. cycle	Initial weight (g)	Final weight (g)	Friction path (m)	Test time (s)	Weight wear (g)	Average wear rate ($g \cdot s^{-1}$)	Average wear rate ($g \cdot m^{-1}$)
Load 60 H							
1	29.958	29.95	9.39	30	0.008	0.00027	0.00085
2	29.95	29.942	9.39	30	0.008	0.00027	0.00085
3	29.942	29.937	9.39	30	0.005	0.00017	0.00053
Load 160 H							
4	29.937	29.835	9.39	30	0.102	0.00340	0.01086
5	29.835	29.667	9.39	30	0.168	0.00560	0.01789

As in the tests of sample No. 2 the wear of the surface layer enriched with mineral particles of sample No. 3 occurred at the first stage at a load of 60 N. Therefore wear at the second stage at a load of 160 N fell on the transition zone and the base metal of the sample. The average weight wear at a load of 60 N was 7 mg.

Table 4 presents summary data on the wear characteristics of the investigated surface layers of samples No. 1-3 and surface layers obtained by laser cladding of alloys of different compositions [15-17] made by us earlier on this installation.

Table 4. Comparative data of wear characteristics of various materials.

Layer material	Wear pressure. (MPa)	Weight wear (mg)	Average wear rate ($mg \cdot s^{-1}$)	Average wear rate ($mg \cdot m^{-1}$)
Sample No.1	0.2	4.3	0.14	0.46
(mineral coating)	0.5	20.0	0.6	1.8
Sample No.2 (mineral coating)	0.2	10.3	0.34	1.10
Sample No.3 (mineral coating)	0.2	7.0	0.23	0.75
Martensitic stainless steel SS 410L	0.5	40	0.67	2.13
Austenitic stainless steel SS 316L	0.5	170	2.83	9.05
Cobalt alloy 2537-10	0.5	13	0.22	0.69
Tool steel M2	0.5	4	0.07	0.21
Tin bronze CuSn11	0.5	120	2.00	6.39
Nickel alloy 1560-00	0.5	4.5	0.08	0.24

As can be seen from the presented data the nickel-base alloy and tool steel have the highest wear resistance which show the minimum values of weight wear at a pressure of 0.5 MPa on the wear surface in the presented range of materials. Sample No. 1 shows comparable toughness to cobalt alloy. The wear resistance of the surface layer of sample No. 3 is approximately two times worse than that of sample No. 1. The least resistance of the three samples under study has the surface of sample No. 2. the wear of which is comparable to the wear of the surface of martensitic stainless steel. However the use of certain methods of protecting parts from abrasive wear is determined along with the physical and technical characteristics of the layers by the economic feasibility in specific operating conditions.

4. Conclusion

A comparative study of the surface of three different samples (steel 20Cr13) modified with ultradispersed particles of various minerals was carried out under conditions of abrasive wear when rubbing against fixed particles under the action of a static load of 60 N and 160 N. According to the results of measurements at a pressure on the wear surface of 0.5 MPa of the three samples studied the highest resistance to wear abrasive action is possessed by sample No. 1 which shows an indicator of weight wear comparable to a coating made of cobalt alloy created by laser cladding and is inferior in this indicator approximately four times to coatings made of a nickel alloy created by laser cladding. and tool steel. The wear resistance of the surface layer of sample No. 3 is approximately two times worse than that of sample No. 1. The least resistance of the three samples under study has the surface of sample No. 2 the wear of which is comparable to the wear of the surface of martensitic stainless steel.

References

[1] Dotsenko A I and Buyanovsky I A 2014 *Basics of tribotechnics: textbook for universities* (Moscow: Infra-M)
[2] Arzamasov V B and Cherepakhina A A 2007 *Materials science and technology of structural materials* (Moscow: Academy)
[3] Bhushan B 2000 *Modern Tribology Handbook. Two Volume Set* (USA: CRC Press Inc)
[4] Kislov S V, Kislov V G, Skazochkin A V, Bondarenko G G and Tikhonov A N 2015 Effective mineral coatings for hardening the surface of metallic materials *J. Russian Metallurgy (Metally)* **7** pp 558 – 64
[5] Skazochkin A V, Bondarenko G G and Kislov S V 2018 Research of Tribological Features of Steel Surface by Creating Mineral Coatings *J. of Engineering Science and Technology Review* 11 **6** pp 138-43
[6] Kislov S V, Kislov V G, Balasch P V, Skazochkin A V, Bondarenko G G and Tikhonov A N 2016 *IOP Conf. Series: Mater. Sci. Eng.* **110** 012096
[7] Skazochkin A V, Useinov A S and Kislov S V 2018 Surface hardening of titanium alloy by minerals *Letters on Materials* 8 **1** pp 81-7
[8] Skazochkin A V, Bondarenko G G and Zukowski P 2019 Research of Surface Wear Resistance of Aluminum Alloy Modified with Minerals using Sclerometry Method *Devices and Methods of Measurements* 10 **3** pp 263–270
[9] Khodabakhshi F, Simchi A and Kokabi A H 2017 Surface modifications of an aluminum-magnesium alloy through reactive stir friction processing with titanium oxide nanoparticles for enhanced sliding wear resistance *Surface and Coatings Technology* **309** pp 114-23
[10] Adebisi A A, Maleque M A and Rahman Md M 2011 Metal matrix composite brake rotor: historical development and product life cycle analysis *Int. J. of Automotive and Mechanical Engineering* **4** pp 471-80
[11] Sheinman E L 2005 Abrasive wear. Review of the American press *J. Friction and wear* 26 **1** pp 100-11
[12] Khrushchev M M and Babichev M A 1970 *Abrasive wear* (Moscow: Nauka)
[13] Stachowiak G W and Stachowiak G B 2004 Wear mechanism in ball-cratering test with large abrasive particles *Wear* 256 **6** pp 600-07
[14] Kumar K and Davim J P 2018 Composites and Advanced Materials for Industrial Applications (Hershey, USA: IGI Global)
[15] Montealegre M A, Castro G, Rey P, Arias J L, Vázquez P and Gonzál M 2010 Surface Treatments by Laser Technology *Contemporary Materials* I–1 pp 19-30
[16] Grigoryants A G, Misyurov A I, Tretyakov R S and Stavertiy A Ya 2012 Comparison of the performance characteristics of coatings made of stainless steels of austenitic and martensitic classes obtained by laser gas-powder surfacing *J. Science and Education* **6** pp 15-24
[17] Dolgovechny A V, Demidova L A and Morozov E A 2012 Technologies for surfacing alloy steel on a carbon steel base *Bull. of the Samara Scientific Center of the Russian Academy of Sciences* 14 **1** pp 550-53

IOP Conf. Series: Materials Science and Engineering **996** (2020) 012024 doi:10.1088/1757-899X/996/1/012024

Prospects for the use of high-speed ESP for oil production

N I Smirnov[1] and E S Skvortsova[2]

[1] Senior Researcher, Mechanical Engineering Research Institute of the Russian Academy of Sciences, Bardina, 4 Moscow, Russia
[2] Educational master Engineering Academy, Peoples' Friendship University of Russia, Podolskoe shosse 8k5, Moscow, Russia

Smir1947@yandex.ru

Abstract. Increasing the ESP speed allows for a reduction in the linear dimensions of the units and the number of standard sizes. By reducing the length, wear resistant materials can be used in each pump stage. However, with an increase in the rotational speed to 10000 rpm, the head of the pumping stage increases significantly (> 35m) and the processes of friction and wear intensify. Test benches for wear testing of pump stages and thrust bearings of hydraulic protectors and submersible electric motors with a rotation speed of up to 12000 rpm have been developed. It was revealed that the greatest change in mass is observed in the guide vane due to erosive wear, which varies depending on the rotational speed in a power-law dependence with an exponent of ~ 3.4. When testing a thrust bearing made of SiC with a rotational speed of 10800 rpm and an axial force of 6000N, the friction coefficient varied in the range 0.028 ... 0.008. Bearing made of hard alloy WC-8Co scored at 4.92MPa load. It is necessary to determine the optimal area of application of high-speed ESPs and the allowable wear values of the pump stages.

1. Introduction

Increasing the rotational speed of electric submersible pump (ESP) installations is a promising direction in the development of mechanized oil production ESP [1]. However, along with the obvious advantages, one should expect problems with reliability due to the intensification of tribological processes and wear of parts. The pump stage is an important component of the ESP pump, the reliable operation of which mainly determines the prospects for the use of high-speed installations. The existing calculation methods [2, 3] are based on the kinetic model of wear and are used only to assess the nature of the flow of liquid with abrasive in the channels of pump stages (PS). Therefore, comprehensive studies of tribological processes in the main tribo-couplings are necessary: thrust bearings, pump stages, radial couplings and seals.

2. General provisions

The head of the pumping stage with an increase in the speed of rotation increases in accordance with the dependence $H_1=H_2\left(\frac{n_1}{n_2}\right)^2$, H_1, H_2 – head pressure, m; n_1, n_2 – rotation frequency, rpm. Accordingly, the length of the pump section $L_1 \sim H_2\left(\frac{n_1}{n_2}\right)^2$. The linear dimensions are reduced mainly for the pump rotor, table 1. For example, the impeller diameter is reduced by ~ 1.5 times. The guide device practically does not change the diametrical dimensions. The diametrical dimensions of radial seals and bearings are reduced in approximately the same proportion.

Table 1. Comparative dimensions of the pump stage of conventional and high-speed pumps.

Main characteristic	Rotation frequency, rpm 3000	10000 [1]
The weight impeller, kg	0.154	0.064
The weight GD, kg	0.275	0.442
Shaft diameter, m	0.017	0.012
D_{MAX} impeller, m	0.071	0.055
The height PS, m	0.025	0.023
The distance between the bearings of the module, m	0.5…1.5	0.230

Changing the size and reducing the number of parts allows you to use highly wear-resistant materials in friction pairs.

On the other hand, a significant increase in the rotation frequency and head pressure of the pump stage increases the requirements for the imbalance of rotating parts and gaps due to wear of the radial and axial seals. Increasing the seal gaps increases the flow of fluid through them, which can lead to parametric failure.

Differences are also observed in the dynamics of the compared installations. If a conventional pump, for example, size 5-50, has 10-11 natural oscillation frequencies in the range of 10-13 Hz, then a high-speed pump operates in the «subcritical» region. Its natural frequency corresponds to a rotation frequency of ~ 208...218 Hz. The nominal rotation frequency of the pump shaft is 10000 rpm (~167Hz).

The main types of wear that occur in high-speed pumps are abrasive wear of seal, erosion wear of the flow part, fretting and shaft corrosion. High load thrust bearing parts are prone to cracking. In the flow path cavitation is observed. This is a distinctive feature of high-speed pumps.

3. The results of the tests of pumping stages

To investigate the effect of the rotation frequency on the wear of the pump stages tested serial pump stages of sizes 5-50 and 5-80 made of powder materials in the frequency range 2970...6000 rpm on a stand with a closed loop of liquid movement and with a one-time filling of abrasive. Quartz sand of the F100 dimension with a concentration of 18 $g \cdot l^{-1}$ was used as an abrasive. The liquid flow rate was 2.0 $m^3 \cdot day^{-1}$. Experience time – 360 min.

The change in the mass of pump stage parts during the experiment illustrates a characteristic feature, figure 1. Thus, the wear rate of the impeller is significantly less than the wear rate of the guide device. In the first case, we observe a linear dependence of the wear rate on the rotation frequency. In the second case, there is a power dependence with an exponent of ~ 3.4. When the rotation frequency is doubled, the wear rate of the impellers increased by 2.6 times, and the guide devices increased by 11 times for the studied speed range. When considering the influence of the speed of rotation on wear, it is necessary to keep in mind the flow rate of the liquid, which is functionally dependent on the speed of rotation.

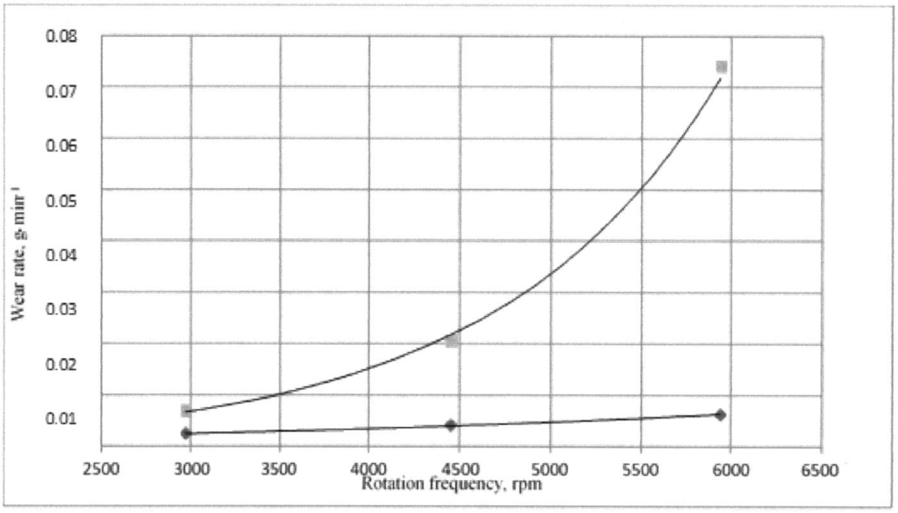

Figure 1. Graph of changes in the wear rate of PS parts, ▪ – guide apparatus, ♦ –impeller.

Based on the analysis of worn surfaces, of measurement of the wear value, it can be concluded that the prevailing share in the total change in the mass of erosive wear is the flow part. Moreover, the guide device is subject to erosion wear to a greater extent than the impeller. It relates to the zones of inhibition of flow in which there are whirls of fluid with abrasive particles at high speed. The impeller has a flow inhibition zone on the input edge of the blade. In the channels of the impeller, the flow of liquid with abrasive particles is directed along the walls. Abrasive particles interact with the surface of the channels at small angles, which is accompanied by a low intensity of erosion wear. It should be borne in mind that the angles of collision of particles with the surface of the flow part change with increasing mass of the abrasive particle.

Tests of the high-speed pumping stage of the AKM pump at a speed of 9500 rpm were carried out on the developed test bench, figure 2, under similar conditions. Test time 5070 min with filling of new portions of abrasive after 600 min It was found that the impeller also wears out less than the guide device. However, the wear rate of the guide device – $1.4 \cdot 10^{-4}$ g·min^{-1} is significantly lower than that of serial PS when tested with a lower speed. This effect can be partially explained by more wear-resistant materials used in high-speed PS (alloy steels, titanium). In addition, the AKM pumping stage was tested at the right boundary of the pressure-flow characteristic.

4. Testing of movable joints PS

One of the main criteria for the performance of pump stages is the amount of wear of movable joints: radial and axial seals. This is especially important for high-pressure pumping stages.

According to the above method, pumping stages made of powder materials of sizes 5-50, 5-80 were tested under the same conditions, but with a changing Pump flow. Analyzing the results, table 2, you can see an increase in the wear rate of radial interfaces V_{rad} and axial interfaces V_{ax} with increasing rotation frequency.

Table 2. Test results for wear of movable joints PS.

Frequency, rpm	Pump flow, $m^3 \cdot hour^{-1}$	$V_{rad} \cdot 10^4$, $mm \cdot min^{-1}$	$V_{ax} \cdot 10^3$, $mm \cdot min^{-1}$	$V_b \cdot 10^5$, $mm \cdot min^{-1}$
2950	1.38	2.6	5.6	1.94
2950	1.58	5.6	4.6	2.9
5705	1.92	15	14.3	3.1
9500[a]	2.2	0.31[a]	-	1.16

[a] PS high speed pump

It is obvious that an increase in feed leads to an increase in the number of abrasive particles per unit time in the wear zone.

The change in the wear rate of pump stage parts is non-linear. The wear rate of the intermediate bearing sleeve V_b on which the shaft end rests is practically proportional to the rotational speed.

Tests of the pump stage of the AKM pump on a high-speed stand, figure 2, showed higher wear resistance values of the intermediate bearing sleeve, table 2, than that of a standard pump stage.

At the same time, the wear resistance of the protective shaft bushings made of hard alloy WC-8Co of this pump stage is slightly lower – $3.1 \cdot 10^{-5}$ mm·min^{-1} than that of the intermediate bearing bushings, but significantly higher than the wear resistance of the serial pump stage. This is due to the bushing's materials - the pump stage AKM they are WC-8Co of solid alloy. An unambiguous conclusion in favor of the higher wear resistance of the moving couplings of a high-speed pump cannot be made only based on these results. The permissible wear value of the mates is of fundamental importance. Both the degradation of the pressure-flow characteristic and the dynamics of the installation depend on it. Apparently, for high-speed installations, this value does not exceed a few tenths of a millimeter. Serial installations can be operated with a wear value of several millimeters without the occurrence of parametric failure.

At the present stage of application of high-speed pumps AKM or other models, the limiting values of the wear of radial seals are unknown, at which a parametric failure of the installation occurs due to a decrease in pressure due to wear.

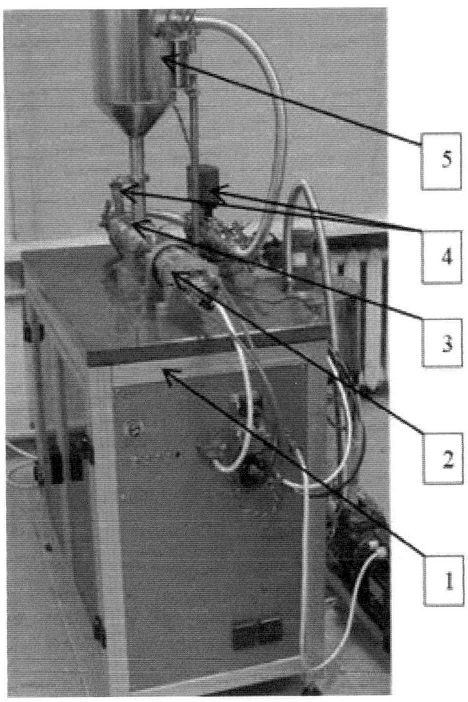

Figure 2. General view of the test bench for pumping stages: 1 – body, 2 – high–speed spindle, 3 – test chamber, 4 – sensors, 5 – mixer, (without a computer control and measurement system).

5. Thrust bearing testing

Structurally, each high-speed pump module has a thrust bearing and two radial bearings. Moreover, due to the reduction in radial dimensions, the thrust bearing has a small contact area compared to bearings of serial pumps, within ~ $5 \cdot 10^{-4}$ m^2. Hard alloy and silicon carbide are used as materials.

We tested a thrust bearing SiC - SiC with diametric dimensions: D = *31* mm, d = 18 mm on the developed stand [4], figure 3, under the conditions: MDPS oil , axial force up to 6000 N, rotation frequency 10800 rpm, temperature 80…85 °C.

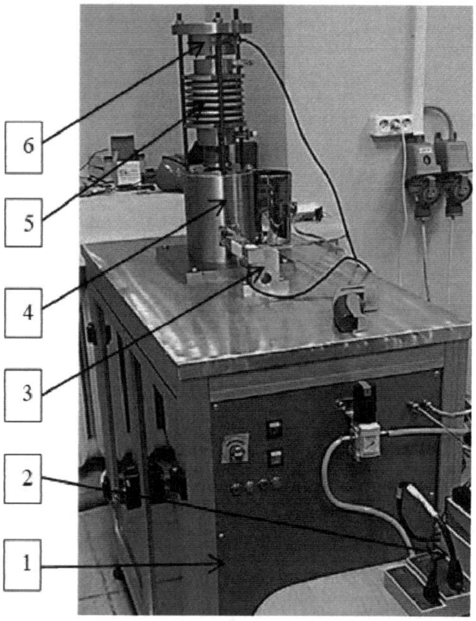

Figure 3. General view of the stand for testing thrust bearings of high–speed pumps: 1 – body, 2 – computer measurement system, 3 – torque generator, 4 – test chamber, 5 – loading system, 6 – axial force sensor.

As a result, the following results were obtained, figure 4. The friction coefficient varies from 0.028 to 0.008 when the axial force changes from 1000 to 6000 N. The maximum pressure value is 12 MPa. Radial cracks have formed on the surface of the silicon carbide rings. The test process at high values of axial force was accompanied by intense heat release. A bearing with similar dimensions made of WC-8Co hard alloy at a pressure of 4.92 MPa suffered a scoring, figure 5.

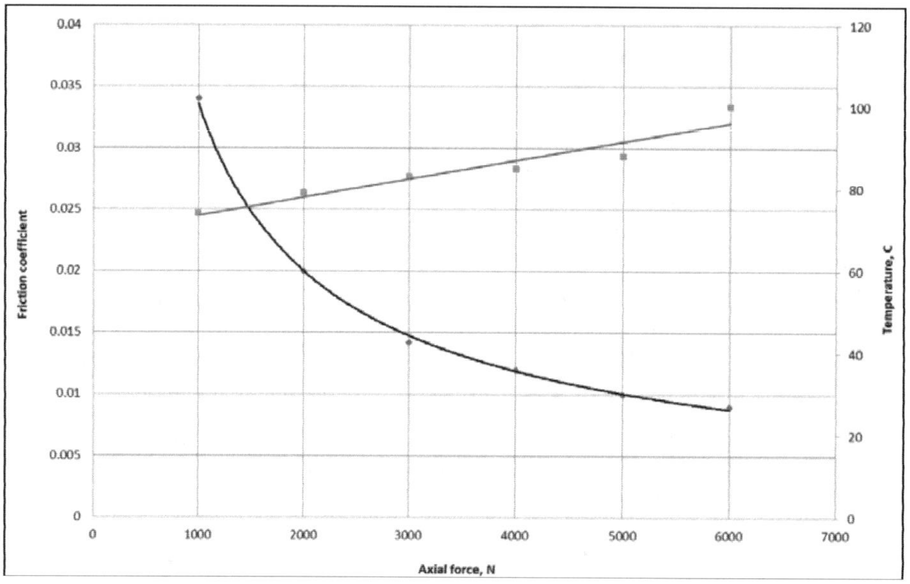

Figure 4. Graph of the coefficient of friction and temperature, ● – coefficient of friction, ■ – temperature.

Apparently, the ultimate pressure that a thrust bearing of this design can reliably withstand is <4 MPa.

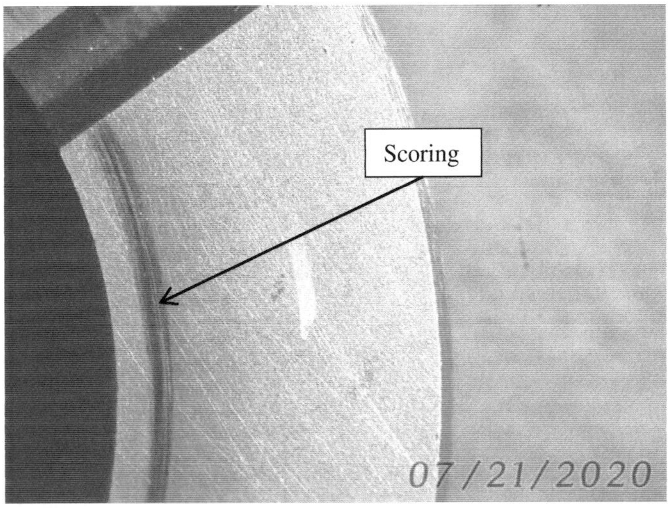

Figure 5. Scoring of the bearing friction surface made of WC-8Co.

The conditions for using thrust bearings must be considered. Their operation takes place in a formation fluid with an abrasive. Therefore, the results obtained can be used for thrusting bearings of submersible electric motors. The thrust bearings of the pump modules require abrasive testing. It can be assumed that the obtained values of the axial force in this case will be even less.

6. Conclusions

1. The prevailing wear mechanisms of high-speed installations of ESP are: abrasive wear of movable joints and erosion (cavitation) wear of the flow path of the pump stage, as well as seizure and thermal scoring of thrust bearings.
2. There is no dependence of the change in pressure on the degree of wear of radial and axial seals, which does not allow determining the optimal field of application of high-speed pumps in terms of complicating factors.
3. Prospects for the use of high-speed electric submersible vane pump installations largely depend on a comprehensive solution to reliability problems.

References

[1] Advertising materials of the company «Lepse» https://www.lepse.com/products/159/
[2] Ostrovsky V G 2013 Management of secondary flows in the stages of oil pumps to reduce their hydroabrasive wear *Ph.D.candidates* (Perm) p 100
[3] Litvinenko K V, Zdolnik S E and Mikhailov V G 2014 Modeling the process of deterioration of characteristics of an electric centrifugal pump under conditions of intense erosive wear *Oil industry* 12 pp 132-135
[4] Smirnov N I and Smirnov N N Development of a set of test benches for high-speed centrifugal pumps *Scientific works of the VI-th int. conf. "Fundamental research and innovative technologies in mechanical engineering" 2019*

Acknowledgment

The publication was prepared wish the support of the "RUDN University program 5-100"

IOP Conf. Series: Materials Science and Engineering **996** (2020) 012025 doi:10.1088/1757-899X/996/1/012025

Tribological studies of carbon seal materials at cryogenic temperatures

N I Smirnov[1], M V Prozhega[1], N N Smirnov[1], A V Petrov[2] and D N Chernenko[2]

[1] Mechanical Engineering Research Institute of the Russian Academy of Sciences, Moscow, 4 Maly Kharitonyevsky Pereulok, 101000, Russian Federation
[2] Stock Company «Research Institute of Graphite – Based Materials «NIIgrafit», Moscow, Elektrodnay, 2, 111524, Russian Federation

prmaksim@gmail.com

Abstract. The results of the tribological properties studies of composite materials at cryogenic temperatures are presented in the present paper. Test equipment has been developed, friction and wear tests have been carried out at a temperature of T = - 196 °C. The test results showed a significant dependence of the composite material wear rate on the load.

1. Introduction

Composite materials based on epoxy resins are applied in details of tribo-couplings due to the good adhesion of epoxy polymers to metals and other materials, high mechanical strength, low shrinkage and water absorption. The most widely used composite materials in Russia are AMS-1, AMS-3, AMS-5M, which feature higher mechanical strength, wear resistance, heat resistance and low coefficient of friction [1]. These materials are used to manufacture piston rings of compressors operating without lubrication, mechanical seals, sliding bearings for dry friction units with normal humidity at elevated temperatures, blades of rotary air pumps, parts of turbo pump units. Composite materials have proven themselves to be good when working at cryogenic temperatures [2]. To increase the resource of devices operating in cryogenic conditions, it is necessary to increase the wear resistance of resource-determining units, which are made of composite materials.

The purpose of this work is to assess the wear resistance of composite materials when operating in cryogenic conditions. To achieve this goal, it was necessary to develop test equipment, elaborate the test methodology and carry out experiments on friction and wear.

2. Materials and test procedure

2.1. Samples for testing

An antifriction self-lubricating composite material based on carbon fabric with epoxy resin as a binder (analogue of AMS-5M) was used as a test object. The counter sample was made of chrome-plated steel 07Kh21G7AKh5 (C< 0.07%; Si < 0.7%; Mn = 6…7.5%; Ni = 5…6%; Cr=19.5…21%; N=0.15…0.25%) with chrome coating or from a WC/Co8 hard alloy. The kinematic scheme "ring - ring", figures 1, 2, was used for the tests.

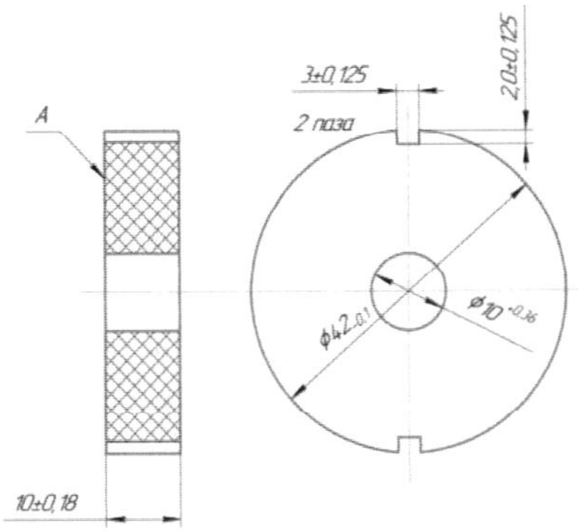

Figure 1. Sketches of composite material sample for testing.

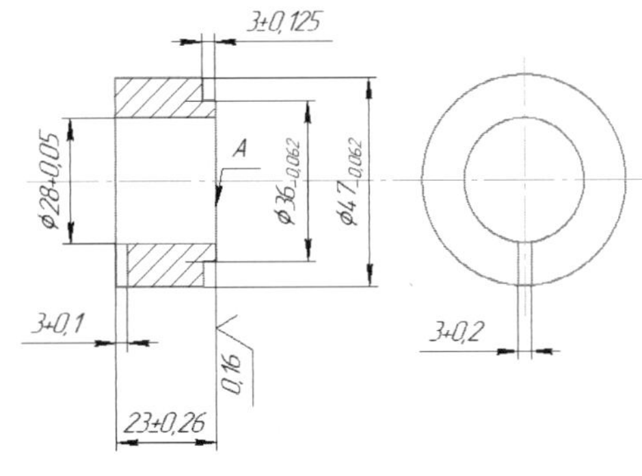

Figure 2. Sketches of chrome-plated steel sample for testing.

2.2. Test equipment

To carry out tests on friction and wear under cryogenic temperatures, a new stand for experiments based on the UMT friction machine was developed at IMASH RAN (Mechanical Engineering Research Institute of the Russian Academy of Sciences), figure 3. On the bed of the UMT friction machine are mounted:

- cryogenic chamber;
- drive unit with a frequency converter;
- support-loading unit;
- sensors for measuring friction force, axial force, temperature and wear.

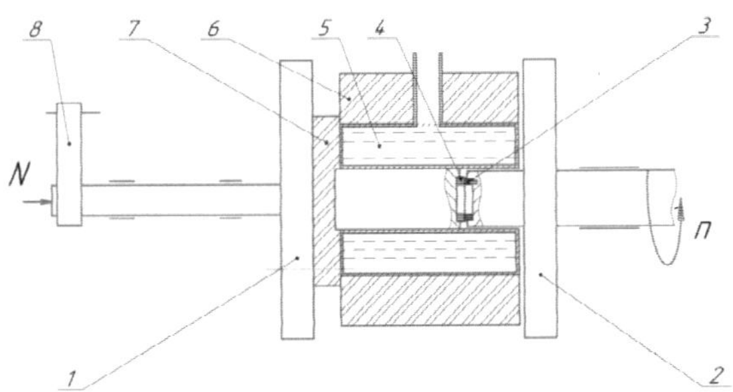

Figure 3. Kinematic diagram of the test unit (1 - support-loading unit; 2 - drive unit; 3 - carbon fiber ring; 4 - steel ring (counter sample); 5 - container with liquid nitrogen; 6,7 - thermal insulation; 8 - meter of friction force; N - axial force; n - rotation frequency).

The axial force N was measured using an M50 membrane compression strain gauge (manufactured by Tenzo-M), the friction force was determined by a T24A beam-type force transducer (Tenzo-M). To measure wear during the experiment, an inductive sensor IWRM 08I9501 manufactured by Baumer was used. The temperature was measured with a Pt100 thermal resistance sensor. To collect data from the sensors, we used an NI DAQ-9184 chassis with an NI DAQ-9219 module. The program for data processing was written in the NI LabVIEW 2017 graphical environment. The post-processing of the data was carried out in the NI DIAdem 2017 program. During the tests, the temperature in the chamber, axial force, coefficient of friction, and wear were continuously recorded.

2.3. Experimental technique

After assembling the test unit, liquid nitrogen was poured into the neck until the moment of moderate evaporation and the temperature of the samples T = - 196°C, figure 4. During the experiment, liquid nitrogen was added. After cooling the samples, the drive unit was switched on, which ensured rotation of the sample with a given frequency, and the samples were loaded with an axial force N of a given value. The loading was provided by excess air pressure in the pneumatic piston by means of a pressure regulator. Built into the disk of the support-loading unit 2 an inductive displacement sensor recorded the wear of the sample during the test, figure 4.

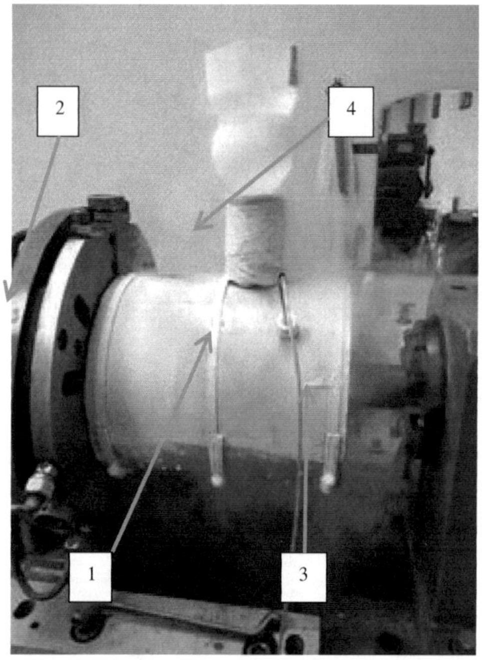

Figure 4. General view of the friction machine test unit (1 - temperature sensor; 2 - disk of the support-loading unit; 3 - shaft of the drive unit; 4 - neck for filling liquid nitrogen).

Loading conditions:

- sliding speed V - 1 $m \cdot s^{-1}$;
- contact pressure p <10 MPa;
- ring temperature T = - 196 ... + 25°C;
- test environment - air atmosphere.

3. Results

Tests at high levels of contact stresses (p = 10 MPa) showed that there is intense wear of the composite material. In addition, the hard chrome coating on the counter sample wears out to the substrate, figure 5, 6, 7.

Figure 5. Wear track of composite sample.

Figure 6. Wear track of the chrome-plated steel.

Subsequently, the value of the load was reduced, and rings made of WC/Co8 hard alloy were used as a counter-sample. The working surface was polished and lapped before each experiment.

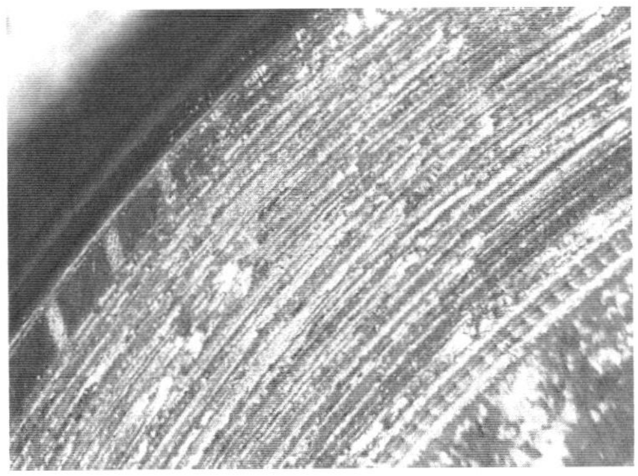

Figure 7. Worn surface of the chrome-plated steel sample (chrome plating).

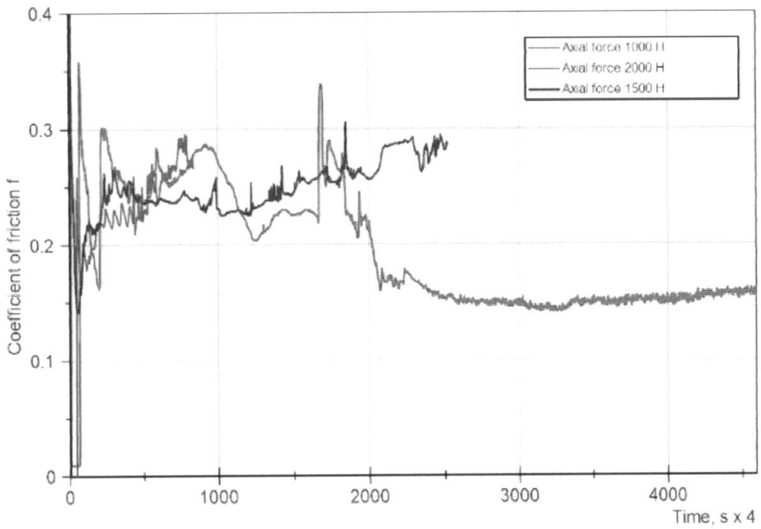

Figure 8. Dependence of the friction coefficient on the load.

A stable value of the friction coefficient f = 0.15 after the running-in stage was observed only at the value of the axial force N = 1000 N. At a higher load, the nature of the friction coefficient graph is unstable, and the value fluctuates in the range f = 0.2...0.3. As a result of the experiments, the dependence of the friction coefficient on the value of the axial load was obtained, figure 8.

Figure 9 shows the wear rate of the coupling when testing the Composite-Chrome coating pair. It can be seen that the wear resistance of the material significantly (around 22 times) decreases with an increase in the axial force from 1000 N to 2000 N. During the tests, it was found that wear changes step by step. Figure 10 shows a record of pair wear at an 1500 N axial force. The graph shows recurring flatter and steeper sections, which indicate different wear rates at the boundaries between the carbon fiber layers. The steeper parts of the wear curve are accompanied by a slight increase in frictional force.

IOP Conf. Series: Materials Science and Engineering **996** (2020) 012025 doi:10.1088/1757-899X/996/1/012025

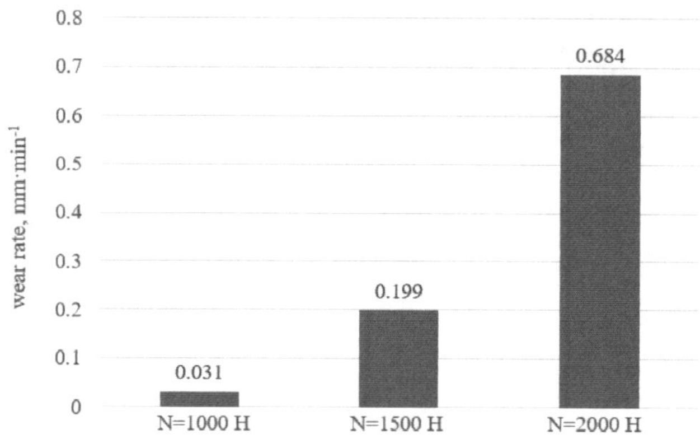

Figure 9. Wear rate of coupling when testing a composite paired with a chrome coating.

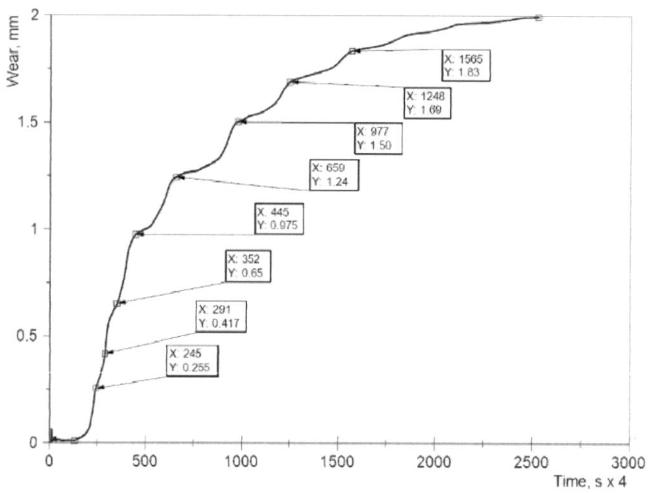

Figure 10. Wear of coupling when testing a composite paired with a chrome coating (axial force 1500 N).

4. Conclusions

1. A stand for testing based on the UMT friction machine has been developed, which allows reliable testing of materials at a temperature of T = - 196°C.
2. The investigated material is satisfactorily efficient up to 2 MPa: wear rate - 0.031 $mm \cdot min^{-1}$; friction coefficient - 0.15.
3. It is noted that the friction force and wear rate change at the boundary of the carbon fabric layers.

References

[1] Tseev N A, Kozelkin V V and Gurov A A 1991 *Materials for dry friction units operating in vacuum: Handbook* (Moscow: Mechanical Engineering) p 192

[2] Hübner W et al Tribological behaviour of materials at cryogenic temperatures 1998 *Wear* **216** pp150-9

Acknowledgments

The authors are grateful to I.S. Splavskiy, a post-graduate student of the IMASH RAN for help with experiments.

Simulation of contact interaction of rough surfaces

V P Tikhomirov and M A Izmerov

The Department of Pipeline Transportation Systems, Bryansk State Technical University, Bryansk, Russia

dm-bgtu@yandex.ru, m.izmerov@yandex.ru

Abstract. Using simulation of contact interaction of rough surfaces, it is possible to determine the dependence of the convergence value on the load applied to the conjugate surfaces, which will allow more accurately solve tribological problems, as well as problems of ensuring a given degree of tightness, etc.

1. Introduction

Existing models of contact interaction are based on assumptions that are used to estimate the parameters of conjugation joint – these are the well-known G–W (Greenwood-Williamson) [1] and M–B (Majumdar-Bhushan) models [2]. G-W model became the basis for calculating the contact interaction parameters of nominally flat rough surfaces. According to G-W model, it was proposed to replace the contact between two rough surfaces with the conjugation of an equivalent rough surface and a solid flat surface without taking into account the interaction of irregularities. Statistical parameters (such as the roughness height, the slope of irregularities, and the rounding radius of the top part of irregularities) are highly dependent on the resolution of the measuring device or any form of signal filter. This circumstance leads to the conclusion that it is necessary to correct the statistical model of contact interaction in order to obtain adequate results. To overcome the drawback of the statistical contact model, Majumdar and Bhushan [1] developed a fractal contact model called M-B model, in which the measure of surface roughness is determined by the fractal dimension D_S ($2<D_S<3$) and the fractal roughness parameter G. A distinctive feature of M-B model is to represent the surface as a fractal object, which is characterized by affine self-similarity. An example of a self-affine fractal object is the surface roughness profile (figure 1).

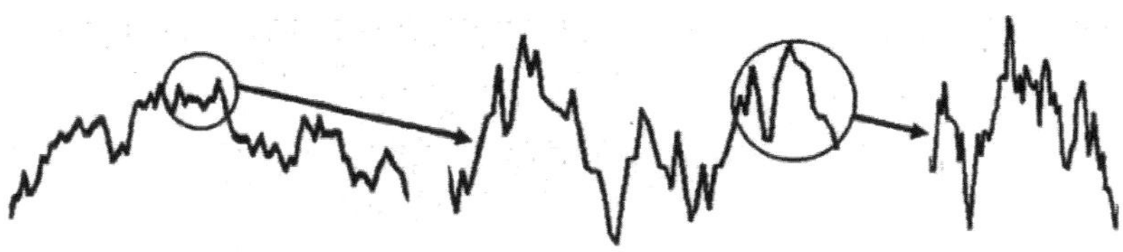

Figure 1. Self-affine surface roughness profile.

As you can see from figure 1, the profile fragments highlighted by circles requires different zoom levels to obtain similarity. Increasing the profile section leads to greater profile detail – revealing the first level of sub-roughness, and then another level. The presence of small details of roughness allowed M-B model to assume that irregularities that come into contact first deform plastically, and then, as the contact area increases, elastically. This is due to the fact that small contact areas, having small radii of curvature, are subjected to plastic deformation. Indeed, the ultimate strain

$$\delta_c = k(H/E)^2 R,$$

where H is the hardness of the less solid interface element; E is the elastic modulus, which depends on radius R of the upper part of asperities. In this case, even at a low level of load on the joint, micro irregularities are deformed plastically until the formed contact area under a certain load becomes elastic (figure 2).

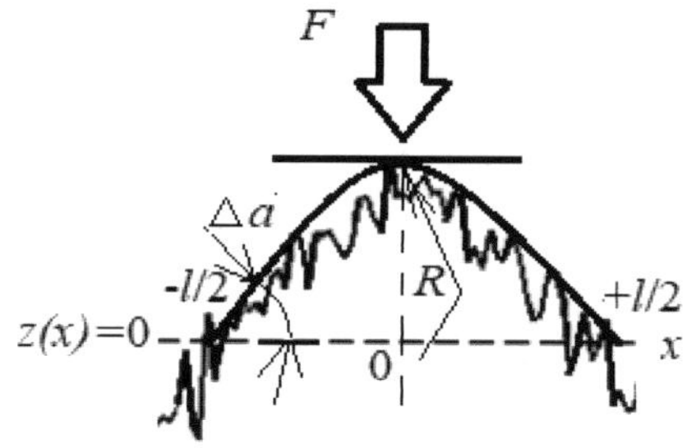

Figure 2. Asperity of micro irregularities.

Here tilt angle of irregularities (the arithmetical average) at the base of a microasperity with size l, is found by the formula

$$\Delta a = \frac{1}{l}\int_0^l \left|\frac{dz}{dx}\right| dx. \tag{1}$$

The profile of an isotropic surface is described by Weierstrass-Mandelbrot equation. Assuming that the base of the asperity (figure 2) is on the midline $z(x)=0$, we use Weierstrass-Mandelbrot equation to describe the shape of the asperity

$$z(x) = G^{(D-1)} l^{(2-D)} \cos(\pi x/l). \tag{2}$$

Here $z(x)$ is the ordinate of an isotropic rough surface measured from the middle plane, G is the fractal roughness parameter, and D is the fractal dimension ($1<D<2$). The presented parameters can be determined (in accordance with [3]) by the following correspondences:

$$G = \left(\frac{R_q\left\{sin\left[\frac{\pi(2D-3)}{2}\right] Gamma(2D-3)\right\}^{\frac{1}{2}}}{L_m^{(2-D)}}\right)^{\frac{1}{D-1}}; \tag{3}$$

$$D = 1.548/R_q^{0.041}.$$

Here R_q is the average square deviation of the profile, and L_m is the basic length of the section when estimating roughness, *Gamma* - gamma function. It is easy to show that for $x = \pm l/2$ we have $z(x)=0$. The maximum height of the asperity from the midline (plane) is equal to $\delta=G^{(D-1)} l^{(2-D)}$.

2. Criterion for the transition from a plastic state to an elastic one

Applying load F to a separate asperity, represented as a set of micro irregularities, leads to the formation of a contact spot, which growth causes the transition to an elastic state. Let asperity deformation w be defined in the following way

$$w = a/(\pi R) \tag{4}$$

Here a is the spot area; R is the radius of the upper asperity part as a fractal object. Equating the loads that cause both the plastic and elastic states of the contact spot, we get

$$a_c H = \frac{4}{3} E R^{1/2} \left(\frac{a_c}{\pi R}\right)^{3/2}. \quad (5)$$

Solving the obtained equation related to the critical value of the contact spot area a_c, we get the following

$$a_c = \frac{9}{16\pi^3} \left(\frac{H}{E}\right)^2 R^2. \quad (6)$$

Using fractal representations (M-B model), let us express the radius of the rounded top of the asperity as

$$R = \frac{a^{D/2}}{G^{D-1}\pi^2}. \quad (7)$$

Substituting the radius value in the expression (6), we find the critical area

$$a_c = \left[\frac{16\pi}{9} \left(\frac{E}{H}\right)^2\right]^{\frac{1}{D-1}} G^2. \quad (8)$$

For example, for a surface with an average square deviation of the profile $R_q = 0.417$ µm, hardness $H = 3000$ MPa, and a reduced elastic modulus $E = 10^5$ MPa (for the basic length $L_m = 800$ µm), fractal dimension of the profile will be $D = 1.605$ and the fractal roughness parameter $G = 8.214 \cdot (10^{-3})$ µm. Then the critical area of the contact spot will be $a_c = 18.89$ µm^2 and the diameter of the contact spot will be d_n=4.9 µm.

3. Micro asperity deformation

According to figure 2 when the fractal asperity is deformed due to small radii of curvature of the sub-roughness, plastic deformation of the sub-rough layer occurs initially until the formed contact area reaches an elastic state. When the load increases, the asperity is deformed elastically, then elastically and plastically and then plastically (figure 3).

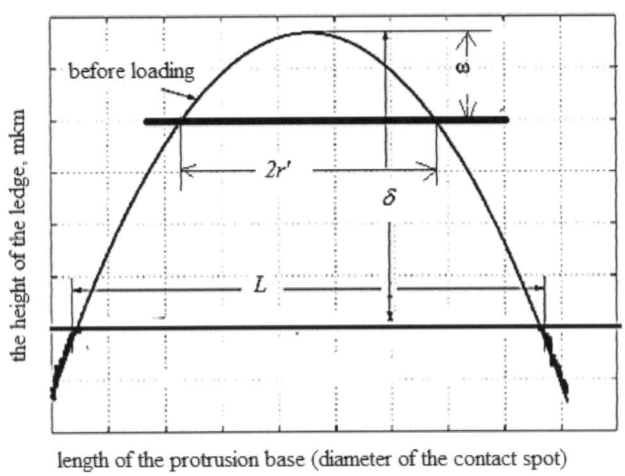

Figure 3. Asperity deformation.

Let us present the relationship between load and deformation for a single micro asperity under different conditions of the contact state.

- The initial state of the loaded contact is characterized by plastic deformation:

$$F_{p-e} = Ha, \quad 0 \leq a \leq a_c. \tag{9}$$

The criterion for transiting from a plastic state to an elastic one is determined by the value of the critical area of the contact spot a_c.

- Elastic contact occurs when

$$\frac{F_e}{F_c} = \left(\frac{a}{a_c}\right)^{3/2}, \quad 1 \leq \frac{a}{a_c} \leq 7.2. \tag{10}$$

here $F_c = \frac{4\sqrt{\pi}}{3} E G^{D-1} a_c^{(3-D)/2}$.

- Elastic-plastic contact, taking into account the data presented in [4], is formed under the following condition

$$\frac{F_{e-p}}{F_c} = 1.302\left(\frac{a}{a_c}\right)^{1.509}, \quad 7.2 \leq \frac{a}{a_c} \leq 206.1. \tag{11}$$

- plastic contact occurs if

$$F_p = H\left(\frac{a}{a_c}\right) a_c, \quad \frac{a}{a_c} > 206.1. \tag{12}$$

4. Multiple contact

Solving the problem of contact interaction of rough surfaces requires knowing the probabilistic distribution of contact spots areas. When surfaces interact under the action of a compressive load, a contact consisting of discrete spots is formed (figure 4). Visualization was performed using the original software developed by the authors.

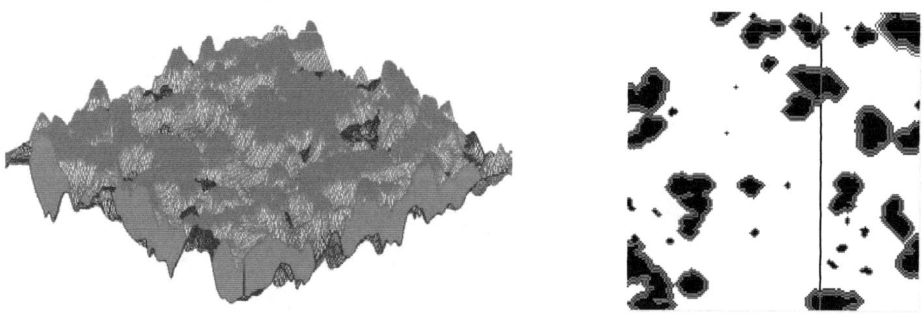

Figure 4. Interface of fractal surfaces.

Normal load growth leads to an increase in the actual contact area, mainly due to the introduction of new irregularities into contact. The fundamental proposition in the theory of contact interaction about the contact discreteness assumes the presence of a certain distribution of the areas of contact spots. It is considered that such a distribution follows a universal power law. For a more accurate estimation of the contact interaction parameters of rough surfaces, the presence of smaller irregularities on the asperity limits the use of the height distribution of asperities. In addition, accounting of such a distribution is difficult for the case when the number of interacting asperities is small.

The probability that a randomly selected spot will have an area greater than a is the following

$$Pr(A > a) = Fa^{-g}. \tag{13}$$

Let us arrange the areas of spots in order of decreasing their size. Selecting one such a spot randomly with a uniform distribution means selecting one sequence number from the list. In this case, we can replace *Pr (A>a)* with *N(A>a)*. Then the total number of contact spots is expressed as the following expression

$$N(A > a) \propto \left(\frac{a}{a_{max}}\right)^{-\frac{D}{2}}. \tag{14}$$

Since the distribution of the areas of the loaded contact is considered relative to the maximum contact area a_{max}, the integral function of the distribution of spots is represented as

$$\mathrm{F}\left(\frac{a}{a_{max}}\right) \propto \left(\frac{a}{a_{max}}\right)^{B}. \tag{15}$$

Despite the structural similarity of the formulas above, their difference is that the number of spots is determined by integral number $N \geq 1$, and the distribution function $F\left(\frac{a}{a_{max}}\right) \in [0.1]$. Let us denote a^* as the relative area equal to the ratio of the spot area to the maximum area of contact $a^* = a/a_{max}$.

Figure 5 shows graphs in logarithmic coordinates of the dependence of contact spots number on corresponding areas and the integral distribution function.

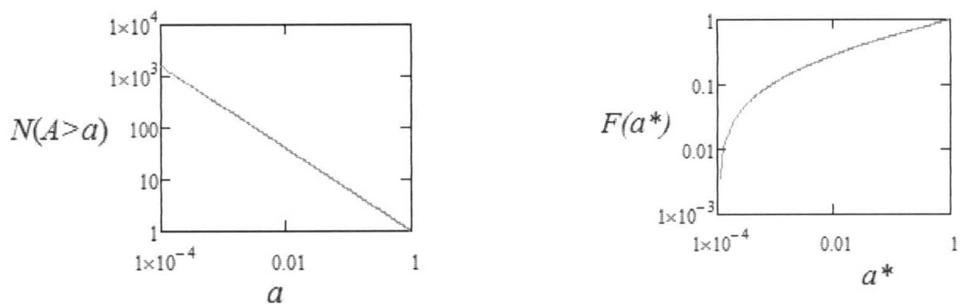

Figure 5. Dependence of contact spots number on contact area ($A>a_{min}$; α=0.8) and the function of spots distribution on the area ($a^*_{min}=10^{-4}$; α=0.8).

Let the distribution density of the relative area of the contact spot corresponds to the power law

$$f(a^*) = C(a^*)^{-\alpha}, \tag{16}$$

where $0 < a^*_{min} \leq a^* \leq 1$, $a^*_{min} = a_{min}/a_{max}$.

Taking the logarithm of the distribution density equation, we can write

$$\log f(a^*) = \log C - \alpha \log a^*. \tag{17}$$

The average size of the contact spot area is

$$\langle a^* \rangle = \int_{a^*_{min}}^{1} C\left(a^{*\prime}\right)^{-\alpha} \cdot a^{*\prime} da^{*\prime}. \tag{18}$$

Here $a^{*\prime}$ – is the the variable of integration.

Taking the integral we will get (при $\alpha \neq 0$)

$$\langle a^* \rangle = \frac{C}{2-\alpha}[1 - (a^*_{min})^{2-\alpha}]. \tag{19}$$

We can find the value of *C* from the normalization condition

$$\int_{a^*_{min}}^{1} C\left(a^{*\prime}\right)^{-\alpha} da^{*\prime} = 1. \quad \rightarrow \quad C = \frac{1-\alpha}{1-\left(a^*_{min}\right)^{1-\alpha}}. \tag{20}$$

IOP Conf. Series: Materials Science and Engineering **996** (2020) 012026 doi:10.1088/1757-899X/996/1/012026

Then the area of the average contact spot (for an infinitely large number of spots, N is the mathematical expectation) is found from the expression

$$\langle a^* \rangle = \frac{1-\alpha}{2-\alpha}\frac{[1-(a^*_{min})^{2-\alpha}]}{\left[1-\left(a^*_{min}\right)^{1-\alpha}\right]}. \tag{21}$$

The integral distribution function is defined by the relation

$$F(a^*) = \int_{a^*_{min}}^{a^*} C\left(a^{*\prime}\right)^{-\alpha} da^{*\prime} = \frac{C}{1-\alpha}[(a^*)^{1-\alpha} - (a^*_{min})^{1-\alpha}] = \frac{(a^*)^{1-\alpha} - (a^*_{min})^{1-\alpha}}{1-\left(a^*_{min}\right)^{1-\alpha}}. \tag{22}$$

Power laws of probability distribution are called distributions with "heavy tails", which are deleted if the normal distribution law is adopted. Power distribution laws, being universal, reflect the structure and processes of complex systems. For simpler systems, exponential distribution laws are used. The power law of distribution in double logarithmic coordinates has the form of a straight line, which indicates scaling behavior, i.e. the absence of selected scales and the presence of fractal behavior.

Admitting that $a^*_{min} = 0,$ we will find

$$F(a^*) = (a^*)^{1-\alpha}. \tag{23}$$

The average area of the contact spot depends on the parameters of the mating surfaces, as well as on the area of the maximum contact spot at a certain level of convergence. In this case, a quantitative estimate of the average area of contact spots can be represented in the following way

$$\langle a \rangle = \frac{1-\alpha}{2-\alpha} a_L, \tag{24}$$

where $a_L = a_{max}$ is the area of the maximum contact spot.

5. Implementation of the simulation algorithm

Using the software developed we can present the results of computer modeling. Figure 6 shows the estimation of contact interaction parameters.

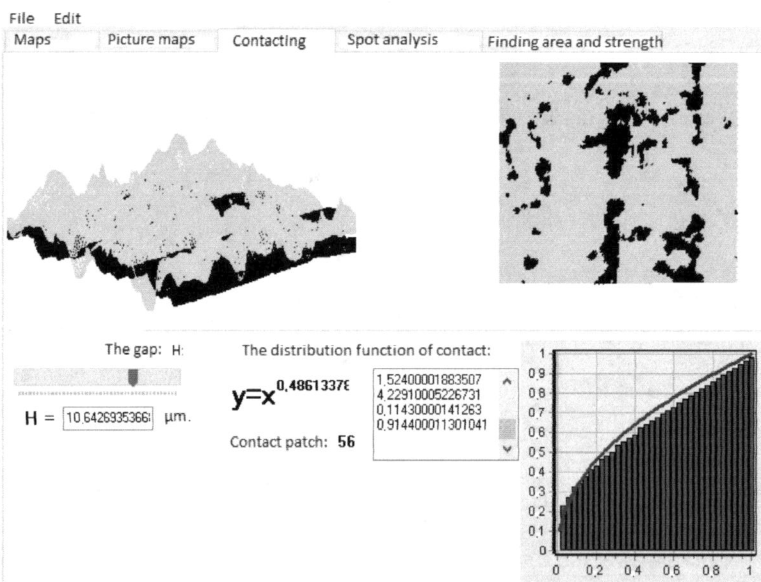

Figure 6. Law of contact spots distribution of milled and polished surfaces.

For example, the interface of milled and polished surfaces (figure 6) gives the law of contact spots distribution $F(a^*) = (a^*)^{1-\alpha} = (a^*)^{0.486}$. Then α = 0.514, and the average area of contact spots will be $\langle a \rangle = 0.327a_L$.

Let us present the simulation procedure as follows.

- Modeling the contact interaction of digital copies of 3D rough surfaces obtained by three-dimensional strip chart recording, or created fractal models that are adequate to the studied surfaces, for example, as presented in the paper [5]. At this stage, we need to get a set of random variables a_i^* -relative areas of contact spots. This can also be done by knowing the law of contact spots distribution for the corresponding pair of conjugate surfaces, for example, using the following dependence (here $x_R \propto Rav[0.1]$is a random variable evenly distributed over the segment from zero to one):

$$a_i^* = x_{Ri}^{1/(1-\alpha)}. \tag{25}$$

- Define to which state each contact spot area should be assigned from the array of areas obtained at the previous stage of modeling, and using the corresponding relationship between the load and the spot area, we find the value of load F_i, which represents the investigated microasperity. In this case, the total load on the conjugate surfaces F_R will be equal to the sum of the forces perceived by the contact spots in all states:

$$F_R = \sum_{i=1}^{N_p} F_p + \sum_{i=1}^{N_{p-e}} F_{p-e} + \sum_{i=1}^{N_e} F_e\,. \tag{26}$$

- If the specified load F applied to the conjugate surfaces exceeds the calculated F_R, this means that the convergence will increase (the load capacity of the rough layer support surface at this convergence is not able to withstand the specified load), and we should increase the convergence and get a new set of contact spot areas. If the specified load is lower, the convergence should be reduced on the contrary. We will repeat the procedure (machine experiment) until the condition (ε is a specified error) is met

$$\frac{|F_M - F|}{F} \le \varepsilon. \tag{27}$$

The initial load at which the area of the maximum spot does not exceed a_c, we can define using the data given in M-B model by the formula

$$F_{p-e} = HA_{r(p-e)} = H\frac{D}{2-D}a_c. \tag{28}$$

For example, using the data given above we can get the following

$$F_{\,p-e} = 3000\frac{1.605}{2-1.605}125.5\cdot 10^{-6} = 1.53\,H. \tag{29}$$

Thus, the initial nominal pressure corresponding to the nominal area Aa = 0.64 mm^2 is equal to p_a = 2.39 MPa.

To test the presented algorithm, the simulation results were compared with the experimental data obtained by the authors in [3], who conducted research using the Leica DCM 3D 3D profilometer. Figure 7 shows a graph comparing the dependence of the contact pressure (in MPa) on the convergence (in μm) for the parameters Rq = 0.417 μm, D_l = 2.6045, G_l = 2.65·10^{-9} m. The upper curve is the model data; the lower curve is the experiment.

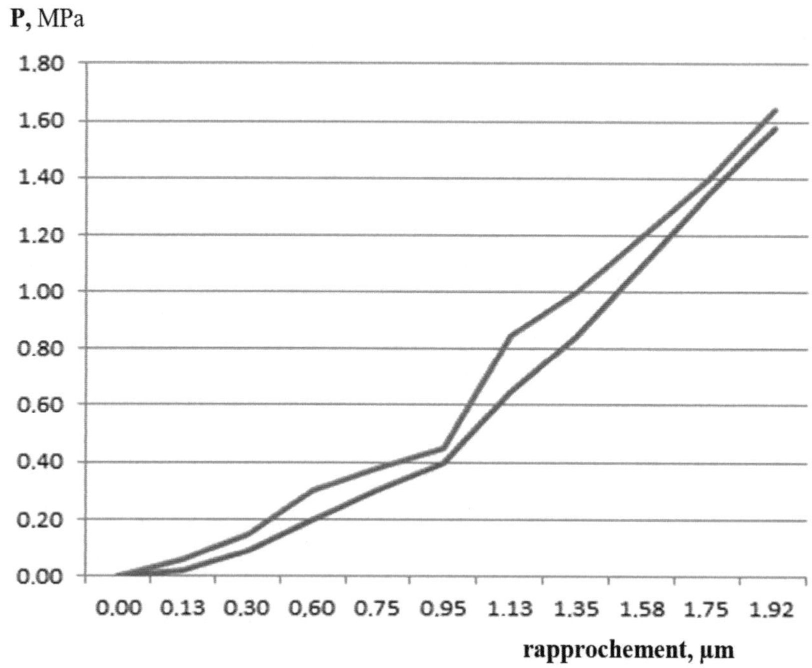

Figure 7. Comparison of experiment and model data.

Thus, it is possible, using simulation modeling of contact interaction of rough surfaces, to find the dependence of the convergence value on the load applied to the conjugate surfaces. The simulation results can be used for solving various tribological problems, evaluating sealing, modeling the behavior of tribosystems under load, etc.

References

[1] Greenwood J A and Williamson J B P 1966 Contact of nominally flat surfaces *Proc. R. Soc. Lond. A* **295** pp 300-19

[2] Majumdar A and Bhushan B 1991 Fractal model of elastic–plastic contact between rough surfaces *J. Tribol.* **113** pp 1-11

[3] Xu K Yuan Y Chen J 2018 The effects of size distribution functions on contact between fractal rough surfaces *AIP ADVANCES 075317* **8** pp 1-14

[4] Kogut L and Etsion I 2002 Elastic-Plastic contact analysis of a Sphere and a Rigid Flat *J. of Applied Mechanics/Trans/ of the ASME* **5** vol 69 pp 657-62

[5] Tikhomirov V P and Izmerov M A 2019 Fractal model of elastoplastic contact of nominally flat rough surfaces *Int. conf. on Aviamechanical engineering and transport (AviaENT 2019): Advances in engineering research (AER)* vol 188 pp 344-50

AUTHOR INDEX

Institute of Physics
Dirac House, Temple Back
Bristol BS1 6BE UK

ISSN: 1757-8981
ISBN 978-1-7138-3303-1

CURRAN ASSOCIATES INC.
proceedings.com